Filter Design for Satellite Communications

Helical Resonator Technology

For a complete listing of titles in the
Artech House Space Applications library,
turn to the back of this book.

Filter Design for Satellite Communications

Helical Resonator Technology

Efstratios Doumanis

George Goussetis

Savvas Kosmopoulos

ARTECH HOUSE

BOSTON | LONDON

artechhouse.com

Library of Congress Cataloging-in-Publication Data
A catalog record for this book is available from the U.S. Library of Congress.

British Library Cataloguing in Publication Data
A catalogue record for this book is available from the British Library.

Cover design by John Gomes

ISBN 13: 978-1-60807-755-7

685 Canton Street
Norwood, MA 02062

10 9 8 7 6 5 4 3 2 1

Contents

Acknowledgments

The authors express their gratitude to Professor Vicente E. Boria Esbert (UPV) for valuable discussions related to multipaction phenomena, and Mark Kunes for his support and valuable discussions on satellite systems and filter technologies for satellites. We would like to thank Professors Ming Yu and Van Dokas for granting us permission to reproduce material on helical filters, and Professor Ming Yu for providing material of his work and valuable discussions.

The authors would like to thank Space Engineering for providing funding for a Ph.D. in helical filters that laid the foundations for this book and for providing material on helical filters.

We thank G. Piccinini (Thales Alenia Space) for the multipaction measurements, and N. Sidiropoulos (Space Engineering S.p.A) for the filter measurements.

We thank Stepan Kucera (Bell Laboratories, Alcatel Lucent) for his help preparing the MATLAB artwork.

Finally, we thank Marco Guglielmi for fruitful discussions on helical resonators.

1

Introduction

Satellites are unique in their ability to provide global coverage. This unique characteristic presents distinctive advantages across a wide range of applications that include the possibility of broadcasting over wide areas, point-to-point communications over long distances, as well as earth observation, surveillance, and navigation. Advances in satellite technology since the early 1960s have enabled a wide variety of services such as those mentioned above for both civil and defence applications. Satellite communication systems play a major role in a broad range of space missions. For broadcasting and communication applications, they deliver the key functionality that provides the required services to the end-users. In other missions, the communication system provides the channel to upload or download information from/to the space segment.

Radio frequency (RF) and microwave filters are essential components in the vast majority of modern wireless communication and radar systems with a number of applications including military, satellite, remote sensing, mobile communication, and many more. Their importance pertains to their capability of frequency-selective transmission/attenuation, which enables transmission of signals in one or more frequency bands (passbands) and reflection of signals at others (stopbands). The filtering characteristics often have a major impact on the overall system performance; consequently, modern wireless systems impose very stringent requirements to the electrical response of microwave filters. It is usually the specific application that dictates the requirements for the electrical performance of the filters. Depending on the application, the electrical performance must also be combined with other features such as high reliability, smaller size, lighter weight, and low-cost mass production.

Over the last few decades, advances in manufacturing technologies, materials, electromagnetic simulators, synthesis techniques, and computer-aided design (CAD) have significantly contributed to major improvements in microwave filters and multiplexing networks.

When it comes to satellite communication systems, the design of RF and microwave filters faces specific challenges associated with many factors, including the host environment, the spectrum scarcity, and the cost of establishing infrastructure in space. Addressing individual requirements of filters for space often necessitates custom technologies and/or conflicting solutions so that a trade-off at component and system level becomes essential. For example, both launching and spectrum allocation costs bring to the forefront requirements for high spectral efficiencies, which in turn require filters with increased power handling capabilities, low dissipation loss, and high selectivity. Although distributed filters comprising of a large number of resonators can be a favorable solution to address these needs, such structures often lead to geometries with dimensions in the order of several wavelengths. As a result, RF and microwave front-end filters are among the bulkiest components in the RF payload—particularly at the lower end of the spectrum. Because the size and mass of on-board equipment have a critical impact on the cost, restricting the footprint while maintaining the performance becomes a significant priority for satellite filter designers. Likewise, the vacuum condition and high temperature variations at the host satellite pose restrictions in the materials that can be used (e.g., to avoid outgassing and/or thermal expansion) and give rise to effects (such as multipactor) that may not be present in high pressure environment. Significantly, ultrareliability for the lifetime of the spacecraft is also imperative. In order to address those challenges, satellite filters encompass a large variety of technologies (e.g., waveguide, dielectric resonators, dual mode waveguide, and dielectric) that cover a wide range of frequencies (from a few MHz up to 40 GHz).

1.1 Helical Resonator Filters

At the lower end of the spectrum, electrical filters are typically realized using lumped RLC elements. At those frequency ranges, the wavelength is much longer than the electrical circuitry, and therefore the lumped element assumption is quite accurate. The performance of RLC filters can be understood considering the frequency dependent reactance of lumped inductors and capacitors [1]. At RF, microwave frequencies, and beyond, the wavelength becomes comparable with the physical dimensions of the circuits, and therefore wave propagation effects should be taken into consideration for a more accurate representation. Even when commercially available, using lumped elements at those higher frequencies is often associated with increased parasitics and dissipation losses that deteriorate the filter performance. Moreover, lumped elements are unsuitable for increased signal power levels. In such cases, distributed filters are employed.

Distributed lowpass filters can be understood in terms of Bragg reflections observed for a wave propagating in a periodic structure. For bandpass filters, the coupled resonator approach may be employed in order to understand their underlying operation. According to this approach, a resonator may be seen as a segment of transmission line between two discontinuities. As a result of the wave reflections at the discontinuities, a standing wave is formed at frequencies where the length of the transmission line is (approximately) equal to an integer number of half wavelengths. At those frequencies, power coming in from one end of the resonator is locally stored in evanescent fields that can then couple to the other end, giving rise to transmission. At frequencies away from resonance, a negligible fraction of the incoming power finds its way into the resonator, giving rise to reflection (stopband) at the input port. Suitable coupling of several resonators in a row can be exploited to a finite transmission band.

Filters employing helical resonators are a class of distributed filters that offers a valuable trade-off between performance and size/mass. They rely on resonators that consist of a coaxial transmission line with a helical core. As a result of the helical geometry, they offer significant miniaturization advantages (typically of one order of magnitude or more) with a relatively small compromise in the electrical performance. They have therefore been primarily used in the lower part of the spectrum and frequencies in the range of few tens of MHz up to about a few GHz. Although the publicly available literature treating the analysis and design of helical resonator filters only covers a few decades [2–4], the lack of rigorous electromagnetic simulators limited those approaches to approximate empirical expressions and/or cumbersome analytical/numerical calculations. The generic design of resonators with helical and folded helical resonators is treated in [4], while the use of helical resonator filters in HF diplexers is presented in [5].

The last fifteen years have seen a renewed interest in helical resonator filter technology. The development of helical resonators for on-board satellite diplexers is presented in [6]. The use of dual-mode helical resonators was proposed in [7], while [8] demonstrates helical filters with cross-coupling between nonadjacent resonators that implement transmission zeros in the stopband. The authors of this book—together with other coauthors—have addressed power handling aspects of helical resonator filters for space [9, 10]. as well as alternative techniques for implementing transmission zeros in the stopband [11]. Although these and other references constitute a significant volume of literature on helical resonator filter technology this is often scattered (with the exception of [4]), and the engineer who wishes to design filters for space or other applications finds this difficult to approach. Moreover, the available literature does not directly account for the significant progress of CAD tools and filter synthesis techniques over the past years in a way that would provide practical guidance on contemporary synthesis methodologies and tools to the design engineer.

1.2 Aim and Contents of This Book

It is the main objective of this book to offer a contemporary and comprehensive approach on the design of RF and microwave filters for satellite applications with emphasis on helical resonator technology. A practical approach that will allow an engineer with basic knowledge on high frequency electronics to understand and use the techniques and methodologies required in modern filter synthesis and design for space as applied to helical resonator filters is adopted. The practical use of modern electromagnetic simulation tools for the analysis and the design of helical resonators and filters in the context of contemporary synthesis strategies is discussed. Examples by means of numerical and experimental results are presented throughout in order to illustrate the concepts and performance characteristics. References are provided to guide the interested reader to more detailed sources. Ultimately, the aim of this book is to provide RF engineers a practical approach that will enable them to confidently engage with the design of helical resonator filters for space and to further extend these techniques to other implementation technologies.

Chapter 2 provides an overview of typical RF communication payload architectures, elaborating on the various roles of the filters in satellite systems. Commencing from the typical transponder architecture and functionality, the specifics of input and output filters and multiplexers are discussed, and a summary of the filters encountered in a communication payload is provided. The key technologies employed for the filter implementation are outlined together with an overview of the corresponding engineering trade-offs. A summary of space qualification procedures and tests for the filter units as defined by common practice in established space agencies is also provided.

Chapter 3 reviews the basic techniques for modern filter synthesis based on the insertion loss method. A practical approach to the basic concepts of microwave filter design that aims to clarify the key concepts and methodologies is adopted. Commencing from the characteristics of the filter response, the design of distributed bandpass filters is introduced from lowpass lumped element prototypes followed by the introduction of immittance inverters and lowpass to bandpass transformations. The realization of impedance inverters by arbitrary discontinuity is presented. The chapter closes with an overview of coupling matrix techniques, which provide increased agility for the advanced design of RF and microwave filters. The main filter synthesis concepts presented in this book provide the basis for more advanced design strategies (not discussed here) and can be extended to other filter technologies.

The operation principles and design strategies for helical resonators are discussed in Chapter 4 in the context of modern electromagnetic software tools and the properties of coaxial transmission lines with helical cores. The adopted approach utilizes the dispersion analysis of helical propagating structures and is illustrated by means of numerical examples. General design guidelines are provided, including a detailed study on the quality factor, which is a critical feature in determining the filter performance. The loading of helical resonators with dielectric (as well as the use tuning elements) are also treated.

Chapter 5 illustrates the application of direct-coupled synthesis techniques for helical resonator filters. Commencing from the synthesis of the coupling matrix (described in Chapter 3), Chapter 5 emphasizes the practical synthesis of the filter's physical dimensions. The external coupling and intercavity coupling between helical resonators is extensively investigated. The physical dimensions of the filter are obtained by matching the coupling matrix values so that a minimum tuning effort is required to finalize the filter. Example designs of high-order filters are presented as illustrations.

Chapter 6 deals with the design of filters with improved stopband performance realized by the introduction of transmission zeros in the stopband in helical resonator technology. The cross-coupling of resonators in folded filter topology through electromagnetic fields (and therefore dispensing the need for additional coupling wires) is presented by means of a fourth-order quasi-elliptic filter example that exhibits two out-of-band attenuation poles by virtue of symmetrical cross-coupling. An alternative technique that exploits in-line coupling of dispersive nature as a means to introduce a transmission zero from a pair of resonators is also introduced and illustrated in detail. Numerical and experimental results are presented to demonstrate the validity.

The implications of utilizing helical resonator filters for high power space applications in terms of the multipactor effect are discussed in Chapter 7. A second-order UHF filter is used as an example to demonstrate the use of European Space Agency's power-handling qualification standards. The analysis shows that these power-handling capabilities are severely limited. A technique for mitigating the power-handling limitations based on the introduction of dielectric covers on the resonators is presented. A prototype employing space qualified dielectric is fabricated and experimentally tested. Measured results demonstrate an improvement in the power handling of the filter with a minimum reduction in the unloaded Q-factor.

Chapter 8 summarizes the work in the previous chapters, draws conclusions, and addresses future challenges pertaining to the helical resonator filters for space.

References

[1] Cameron, R., C. Kudsia, and R. Mansour, *Microwave Filters for Communication Systems: Fundamentals, Design and Applications,* Hoboken, NJ, Wiley-Interscience, 2007.

[2] Macalpine, W., and R. Schildknecht, "Coaxial resonators with helical inner conductor," *Proceedings of the IRE,* Vol. 47, No. 12, pp. 2099–2105, 1959.

[3] Zverev, A., *Handbook of Filter Synthesis,* New York: Wiley, 1967.

[4] Vizmuller, P., *Filters with Helical and Folded Helical Resonators,* Norwood, MA: Artech House, 1987.

[5] Levy, R., and K. K. Andersen, "An optimal low loss HF diplexer using helical resonators," *IEEE MTT-S International Microwave Symposium Digest,* 1992.

[6] Crino, V., and S. Kosmopoulos, "Metop—the SRA diplexer design-technical report," *Space Engineering S.p.A,* Rome, 1999.

[7] Kwok, R., and S. Fiedziuszko, "Dual-mode helical resonators," *IEEE Trans. Microw. Theory Tech.,* Vol. 48, No. 3, pp. 474–477, 2000.

[8] Yu, M., and V. V. Dokas, "High performance helical resonator filters," 34th European Microwave Conference, 2004.

[9] Doumanis, E., G. Goussetis, W. Steffe, D. Maiarelli, and S. Kosmopoulos, "Helical Resonator Filters with Improved Power Handling Capabilities for Space Applications," *IEEE Microwave Components Letters,* Vol. 20, No. 11, pp. 598–600, 2010.

[10] Doumanis, E., G. Goussetis, S. Kosmopoulos, and M. Guglielmi, "Conical-Shaped Helical Resonators for Improved Multipaction-Free Power Thresholds," *European Space Agency Microwave Technology and Techniques Workshop*, Nordwijk, 2012.

[11] E. Doumanis, G. Goussetis, and S. A. Kosmopoulos, "Inline Interdigital Pseudo-Elliptic Helical Resonator Filters," *IEEE Microwave and Wireless Components Letters*, Vol. 21, No. 8, pp. 400–402, 2011.

2

Microwave Filters for Satellite Systems: Requirements and Constraints

2.1 Introduction

This chapter briefly describes the communications payload of a satellite system, with its main emphasis on microwave filter utilization, function, requirements, and constraints. There are several key aspects and conditions unique to satellite communications that impose additional constraints from a system and filter design point of view. For example, several filter specification requirements relate to the unique aspects of (for example) temperature, pressure, and vibration associated with space environment as well as the function that the payload is serving. Some of these requirements may be conflicting and interdependent. These factors mandate both design trade-offs and qualification test protocols. Some key relevant facts are:

- The satellite launch cost per kilogram for 2010 was estimated at approximately €20K ($30K) [1]. It has been reduced down from £30K ($50K) per kilogram a decade ago [2].
- The on-orbit life requirement for satellites has increased from five years in the 1980s to 18 years in 2010 [1].
- Satellite mass has more than tripled, and power available for operating the satellite and its payload has increased tenfold since 1980 [1].

Low-cost, high spectral efficiency—in addition to high reliability and throughput—are key drivers in satellite systems and shape key performance requirements (e.g., low loss, high linearity and high power handling), as well as some of the major constraints (e.g., compact and lightweight) of the on-board satellite electrical equipment. Some key factors related to satellite systems are outlined below.

Competition—low-cost. This is one of the main drivers. It practically affects every aspect of high-power commercial satellites systems. Complete communication satellites are very costly to build and test. Launching cost represents a significant fraction of the entire cost, which is directly related to the overall mass and volume of the satellite. The size and mass of on-board equipment therefore has a critical impact on the cost. As a direct result of the cost constraints, spectral efficiency and quality of service become significant.

High reliability—long life. Satellites operate in a hostile space environment and must work without maintenance for fifteen years or longer (lifetime for a common geostationary satellite can exceed fifteen years). They must be able to survive the high shock and vibration environment experienced during launch and must be able to withstand the vacuum and the extreme temperatures while in orbit. Additional design considerations relate to shielding sensitive electronics from the effects of radiation from the sun and interstellar space. The equipment must be ultrareliable in order to operate in space without failure. Consequently, each subsystem and component must demonstrate that it can perform its function under exposure to all of the extreme environmental and operational stresses that it will encounter in addition to the fifteen (plus margin) years of degradation caused by the unique radiation and thermal conditions in space [1]. In this context, redundancy is also a key element in the operation of such complicated systems. For example, switch matrices are often employed to provide redundant units in case of failure; these, however, bring to fore additional size, weight, and ultimately cost constraints. Finally, when it comes to radio electronics, high stability of frequency versus temperature is necessary and must be achieved with low mass.

Narrowband—low loss. Such constraints often relate to the requirement for efficient spectrum utilization so that the use of the available communication bandwidth and space segment infrastructure are maximized. Those, in turn, necessitate minimal bandwidth wastage in intervening guard bands by placing the various repeater channels close together. Subsequently, narrowband bandpass channel filters with a steep rise of rejection at the band edges for input demultiplexers and output multiplexers are required. This is particularly so because RF/microwave filtering is a key functionality in reducing the noise and hence improving signal-to-noise ratio (SNR) of the signals. For similar reasons, the available bandwidth should provide high-quality transmission, a property necessitating low levels of amplitude and group delay distortion.

Compact and lightweight. Technological development and market drives have over the past few decades significantly increased the mass of a satellite that can be inserted in geosynchronous orbit and brought down the launch cost per kilogram. Even so, launch cost is a large part of an operator's investment, and this drives satellite technology toward miniaturization and higher efficiency in order to provide much larger communications capacity per unit mass. The components must therefore be compact and lightweight. This is one of the key requirements for on-board microwave filters because, in a communications satellite, the mass and physical volume of a transponder's filters and multiplexing networks may represent an appreciable proportion of the allowable payload [3].

High power. Low-loss filters are essential in the transmitter-output circuit in order to reduce thermal management considerations and to optimize the power efficiency of the payload. The latter aspect is significant because power generation in space is costly and transmitter power is ultimately limited by the available power from solar cells. Meanwhile, increasing the signal power levels is critical in providing improved SNR and quality of service. This, in turn, drives the ever-increasing power level of downlink signals. A problem specific to the space environment and high power signals that becomes increasingly relevant as the signal power levels increase is that of multipaction. Multipaction is a secondary emission phenomenon that occurs under high-vacuum conditions and can result in the destructive breakdown of microwave equipment and loss of provided service. RF and microwave filters for space are therefore required to produce low loss and to be able to withstand increasing power levels.

Operating frequencies. Satellite services are covered by a wide range of microwave frequencies. The services provided by each payload determine the frequency of operation of the RF payload. For example, navigation and mobile satellite systems are typically operated in the L (1–2 GHz) and S (2–4 GHz) bands. Remote sensing applications are mostly offered in C (4–8 GHz) band. For commercial communications (due to the increasing demand of high-quality services), the operating frequency band has gradually shifted from the C-band towards the Ku-band (12–18 GHz) and Ka-band (26–40 GHz). A brief overview of the utilization of frequency bands and corresponding services provided by the European Space Agency [4] is reproduced in Table 2.1. A range of filter technologies are available to address the specifics of each operating frequency band.

2.2 System Perspective

It is the task of the payload system engineers to specify the performance parameters of individual equipment in such a way that the required system performance is achieved for the payload as a whole. Key payload electrical parameters of a communication payload are [3]:

Table 2.1
Satellite Frequency Bands by European Space Agency

Band	Frequency (GHz)	Description
L-band	1–2	Global Positioning System (GPS) carriers and satellite mobile phones, such as Iridium; Inmarsat providing communications at sea, land, and air; WorldSpace satellite radio.
S-Band	2–4	Weather radar, surface ship radar, and some communications satellites (especially those of the National Aeronautics and Space Administration, NASA) for communication with ISS and Space Shuttle. In May 2009, Inmarsat and Solaris mobile (a joint venture between Eutelsat and Astra) were awarded each a 2×15 MHz portion of the S-band by the European Commission.
C-Band	4–8	Primarily used for satellite communications, for full-time satellite TV networks, or for raw satellite feeds. Commonly used in areas that are subject to tropical rainfall because it is less susceptible to rain-fade than the Ku-band (the original Telstar satellite had a transponder operating in this band; it was used to relay the first live transatlantic TV signal in 1962).
X-Band	8–12	Primarily used by the military. Used in radar applications including continuous-wave, pulsed, single-polarization, dual-polarization, synthetic aperture radar, and phased arrays. X-band radar frequency subbands are used in civil, military, and government institutions for weather monitoring, air traffic control, maritime vessel traffic control, defence tracking, and vehicle speed detection for law enforcement.
Ku-Band	12–18	Used for satellite communications. In Europe, Ku-band downlink is used from 10.7 GHz to 12.75 GHz for direct broadcast satellite services, such as Astra.
Ka-Band	26–40	Communications satellites, uplink in either the 27.5 GHz and 31 GHz bands, and high-resolution, close-range targeting radars on military aircraft.

- Antenna coverage;
- G/T (the ratio of receive antenna gain to system noise temperature);
- Effective isolated radiated power (EIRP).

Elements of payload system performance analysis that directly reflect performance restraints to microwave filters on board satellites include:

- Linearity;
- Group delay variation;
- Spurious outputs;
- Isolation between channels.

Group delay variation. The group delay refers to the time delay of a modulated waveform signal passing through the communications equipment. The group delay is equal to the slope of the phase versus frequency characteristic [5]. In a nonlinear frequency characteristic (slope varying with frequency), different components of the modulated signal experience different time-delays, which eventually leads to the distortion of the baseband signal. System specifications commonly impose stringent requirements on the group delay variation across the channel bandwidth. In practice, such limitations are mostly relevant to resonant components and therefore are of significant importance to the filters; they are particularly important to the narrowband channel filters (the channel filters of the input and output multiplexers), which are the main source of group delay variation. Group delay distortion in one element of the RF system can be compensated by equalizing circuits, which is more practical to perform at low-signal power levels prior to the signal passing through the main amplification stages prior to retransmission (i.e., the high-power amplifiers (nonlinear devices)).

Spurious products. The payload specification invariably places limits on the levels of unwanted spurious outputs from the transponder. One aspect of payload design is the prediction, and where necessary, the elimination by the inclusion of filters (or by other means) of these emissions that may cause harmful interference to the quality of service for users either of the same system or of other systems. It may even cause problems in the operation of the payload itself. Spurious products have a variety of origins including [3]:

- Spurious mixing products at the frequency converters;
- Spurious local oscillator outputs;
- Intermodulation products;
- Harmonics of the output frequency;
- Power supply ripple components (for instance, at the switching frequency of the voltage regulators).

Although it is often possible to identify (by inspection) the most likely sources of troublesome outputs, there is ultimately little alternative to a systematic examination of each possible unwanted product in order to determine where it may affect the payload system and what harmful effects it may cause.

Passive intermodulation and multipaction. One problem associated with the high-power output circuits and antennas is that of passive intermodulation (PIM). Passive components in the high-power path of the payload may create small but harmful electrical nonlinearities. PIM arises primarily due to the formation of effective diode junctions at the interface between different materials; it can also be the result of dirt, thin oxide layers on metal surfaces, or mechanical imperfections on the joints. Although the signal levels of these frequency components are very weak when compared with the transmitted signals, they may be of comparable power level with the received signal and would therefore cause problems, particularly when manifested at the common paths between transmit and receive (e.g., antenna, cable, or duplexer). PIM products generated in a unit within the transmit band cannot be removed by filtering. The nonlinear behavior of passive components has emerged as an important phenomenon and has become a factor in determining the frequency plan and equipment performance for all high-capacity and high-power satellite systems [6].

PIM depends upon the materials in contact, surface conditions, workmanship standards and subsequent handling of the equipment. All components must be controlled by extreme care in manufacturing (i.e., extreme cleanliness, suitable choice of materials, very precise machining, and high contact pressures between mating surfaces) [3]. The space industry has built up an extensive experience base in the designing, implementing, and measuring of high power and low PIM equipment. A cost-effective approach to low PIM is to follow a set of guidelines throughout the stages of a satellite program. Such guidelines are presented in [2, 6–8]. PIM-related aspects specific to filter technologies are addressed in Section 2.3.

The second problem associated with high power signals is a type of electrical breakdown that occurs only in a high vacuum. Multipaction is caused by acceleration of electrons by an RF field in a gap between metal surfaces. High impact secondary emissions (for combinations of frequency, voltage, and gap size) can produce an electron avalanche. Subsequently, this can cause intermodulation (IM) products, broadband noise, and in extreme cases, physical damage. Again—in practice—the onset of the phenomenon is sensitive to surface finish and contamination. Multipactor is an aspect that should be studied in the development of filters along the downlink chain. Today, multipaction is typically addressed; allowing a generous design margin is a must [3, 6, 9]. Multipactor is discussed in more detail in Chapter 7.

2.2.1 Communication Payload: Filter Functions

The frequency band allocated to a satellite system is commonly split into a number of RF channels. This is due to the practical constraints associated with the nonlinear performance of high power amplifiers [6, 10]. For a lower number of carriers in each subband, the intermodulation noise generated by the channel amplification stage is reduced compared to the noise that would have been generated if the amplifying stage had operated on the total system bandwidth [11]. A generic block diagram of the communications payload of a satellite repeater is shown in Figure 2.1. It includes the low-noise receiver

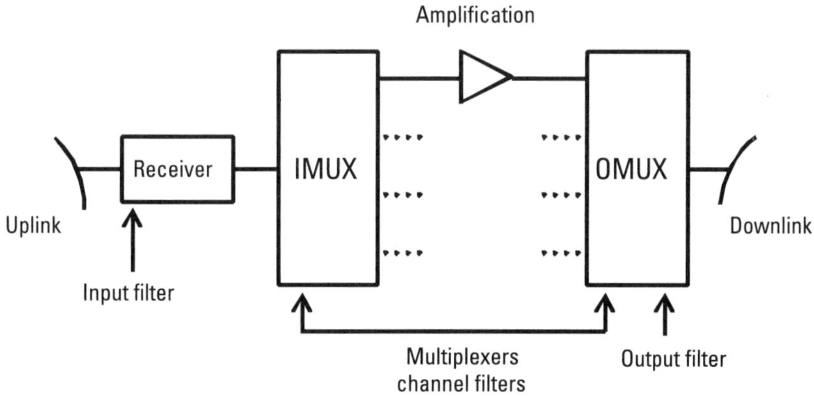

Figure 2.1 Simplified schematic diagram of the communications payload of a satellite [10].

section, the channelizer section, the high-power amplifiers, and the high-power transmitter section [7, 11]. The individual channels are serviced by transponders.[1] Figure 2.2 shows a simplified block diagram of an international satellite payload (C-band: 6/4 GHz), as reproduced from [12]. In these diagrams, the duplexing function is omitted for simplicity.[2]

In the receiver section, a receive antenna is connected to a wideband filter followed by a low-noise receiver. The wideband receive filter is required to ensure that only the useful bandwidth is fed to the low-noise amplifier (LNA) and that all other signals are attenuated. Typically, this is a broadband bandpass filter covering the entire uplink frequency spectrum often followed by a low-pass filter to clean up the received spectrum from out-of-band products received at the (wideband) antenna as well as the out-of-band noise. At the receive antenna, the signal strength is at its lowest level; and therefore it is crucial to minimize the energy loss of this filter. Figure 2.2 demonstrates the concept of redundancy where a redundant wideband receiver is included employing a switch. In a real payload, many functions are duplicated, and complex switch matrices may be employed to provide redundancy in case of failure. Following a down-conversion mixer, a bandstop filter is meant to reject the spurious outputs of the mixer (the bandstop filter is there to reject any spurious products falling in the usable receive and transmit bandwidth).

Demultiplexing is achieved with the signal channelized via an input multiplexing network (IMUX), which is essentially composed of narrow bandpass microwave filters. The purpose is to create channels (subbands) of reduced spectral width that reduce the nonlinear distortion during amplification. Each RF channel is occupied by a number of carriers, so the total number of channels would include all the carriers occupying the system bandwidth. Modern satellite channel filters have a requirement for very rapid roll-off when the channels are closely spaced. The most common approach for the IMUX is the channel dropping technique, which utilizes directional devices (circulators) to eliminate the interaction between the channel filters. Figure 2.2 illustrates the circulator-coupled (channel dropping technique) IMUX technique. Channel filters at the IMUX level have to deliver the high levels of interchannel isolation that are required in order to improve spectral efficiency and relax the performance requirements of the high power output channel filters.

---

1. The word "transponder" is sometimes used instead of "channel" and designates the equipment that operates within a given subband.
2. A duplexer or diplexer is the device that allows the system to utilize a common antenna that receives and transmits simultaneously.

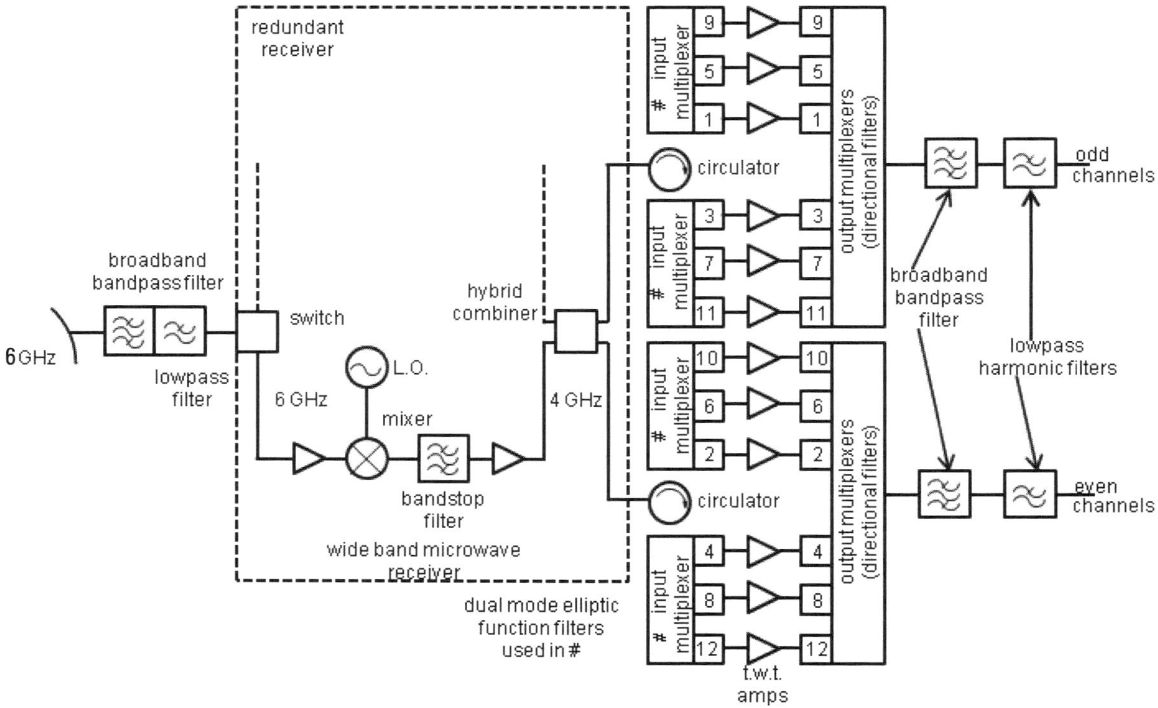

Figure 2.2 Schematic diagram of an international communications satellite payload depicting the redundancy at the receiver, the noncontiguous multiplexing, and the circulator-coupled multiplexing technique. (Copyright 2014 IET [12].)

Two alternative multiplexing schemes have been mostly employed for satellite systems. Early systems employed the noncontiguous multiplexing schemes (shown in Figure 2.2). Organization of the channelized part into even and odd channels left a guard band between each channel with a width equal to that of one channel for each group. This allowed the individual channel filters to be designed separately. A 3-dB hybrid combiner is then connected to the output port of the two even and odd multiplexers. Such a configuration requires an antenna feed network with two input ports [6]. The advantage of this approach is the design simplicity of each multiplexing network due to larger separation between the individual channels. On the other hand, the incoming signal power is shared between the two ports of the antenna, thus reducing the signal power levels in each branch. The alternative is to combine all the transponders in a single device. The complexity of this latter approach can be overcome with significant technology advances over the years.

Following the IMUX, each RF channel is then amplified separately and recombined by an output multiplexer network. The output multiplexer (OMUX) consists of a number of bandpass filters, whose outputs are commonly connected to a common waveguide (manifold). The output multiplexer that follows the high-power amplifier must have low in-band loss (unlike the input multiplexer, where in-band loss is usually less stringent). The high-power output multiplexing network further requires harmonic rejection filters. Two schemes are commonly used. One involves a common filter at the output port of the multiplexer. This approach imposes critical requirements to the power handling capabilities of such filters because they must handle the combined power of all the RF channels. The alternative approach requires individual filters for every channel that dictates higher volume, mass, and increased loss.

Figure 2.3 shows a set of satellite transponders for (a) a C-band and (b) a Ku-band system. The C-band transponder uses a single converter (down converter, D/C) and signal processing at the downlink frequency of 4 GHz, whereas the Ku-band system uses D/C to 1 GHz for signal processing followed by up-conversion (U/C) for the downlink. The Ku system (uplink 14 GHz/downlink 11 GHz) shares many parts, but the downlink frequency (11 GHz) is too high for the elements in each transponder, so the input is mixed down from 14 GHz to 1 GHz for demultiplexing and equalization, then mixed up to 11 GHz for power amplification, frequency multiplexing,and retransmission. This configuration and system architecture

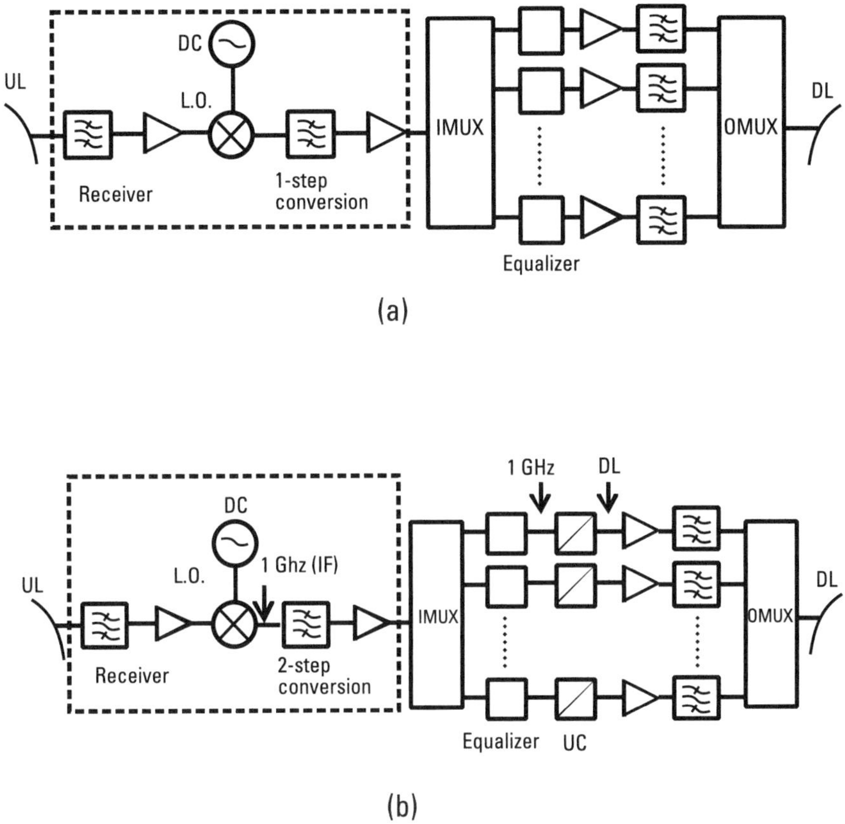

Figure 2.3 Simplified block diagram of a (a) C-Band payload and (b) Ku-band payload.

reflects directly to the specification requirements and frequency planning with direct consequences to the filter requirements.

2.2.1.1 IMUX/OMUX Configurations

In order to achieve the functionality described above, input and output multiplexers are designed based on a number of configurations. Channel filters are readily dependent on the different configurations as summarized below:

Hybrid-coupled multiplexer. Figure 2.4 shows the schematic block diagram of a hybrid-coupled multiplexer. In this multiplexer configuration, the direc-

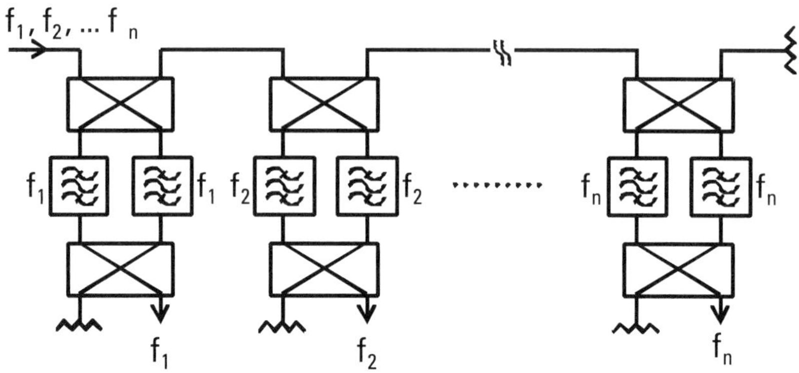

Figure 2.4 Schematic diagram of a hybrid-coupled multiplexer configuration.

tional properties of the hybrid coupler are used to minimize the interaction between channel filters. This allows for the relaxing of the design complexity of the channel filters. Subsequently, this configuration is amenable to a modular concept and a flexible frequency plan, allowing for the the integration of additional channels at a later date without disrupting the existing multiplexer design. The signal of each channel is divided into two paths and recombined at the output. This multiplexer layout requires two identical channel filters and two identical hybrid couplers for each channel. Thus, it has the disadvantage of large size and weight. The signal also undergoes the insertion loss of the two hybrids. The power handling capability of this multiplexer is double the power handling capability of other multiplexer configurations, as only half the input power is fed to each filter. Another design consideration of such multiplexers is the phase deviation between the two filter paths that the two signals undergo before they add constructively at the channel output. The structure, therefore, must be fabricated with tight tolerances to minimize the phase deviation.

Circulator-coupled multiplexer. This type of multiplexing (Figure 2.5)—also known as *channel dropping technique*—uses the directional properties of a circulator [5] in order to isolate the channel filters. This type of multiplexing technique is the most common realization approach for the IMUX [13]. This multiplexer has the disadvantage of higher insertion loss due to the additional loss incurred by subsequent circulators, which depends on the number of sequential channels. Additional couplers are commonly employed to utilize

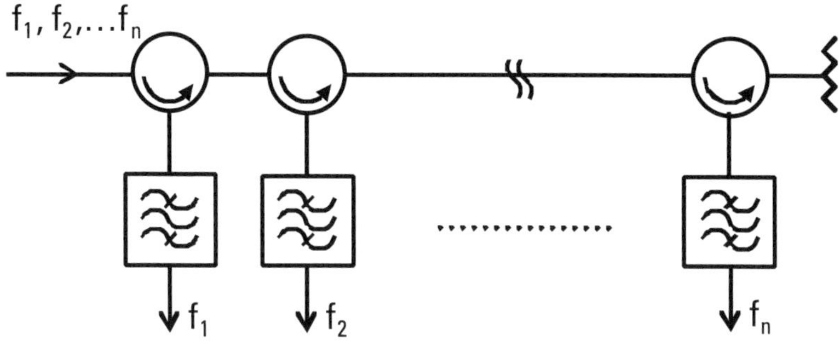

Figure 2.5 Schematic diagram of a circulator-coupled multiplexer configuration.

channel group delay equalizers (IMUX) and isolators are required at the output for matching purposes over a broad bandwidth [2].

Directional filter multiplexer. Figure 2.6 depicts a block diagram of a directional filter multiplexer. A directional filter is a four-port device in which one port is terminated in a load. The other three ports of the directional filter essentially act as a circulator connected to a bandpass filter with no need to use ferrite circulators. The signal incident at one port is bandpass filtered to a second port while the reflected signal from the filter exits the third port towards the rest of the multiplexer circuit. This multiplexer, like the hybrid-coupled and the circulator-coupled configurations, uses directional properties and is therefore amenable to a flexible frequency plan. It is, however, limited to narrowband applications.

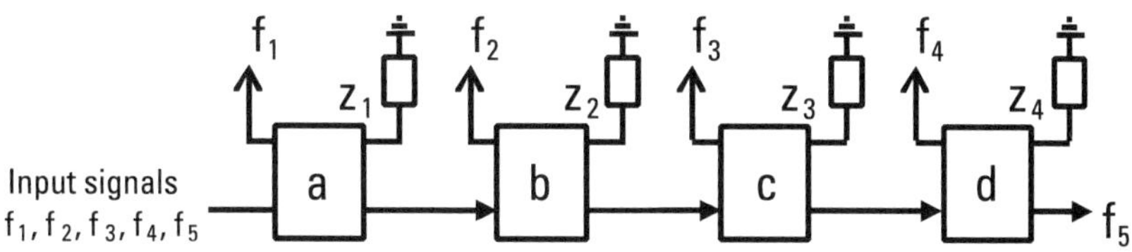

Figure 2.6 Schematic diagram of a directional filter multiplexer configuration.

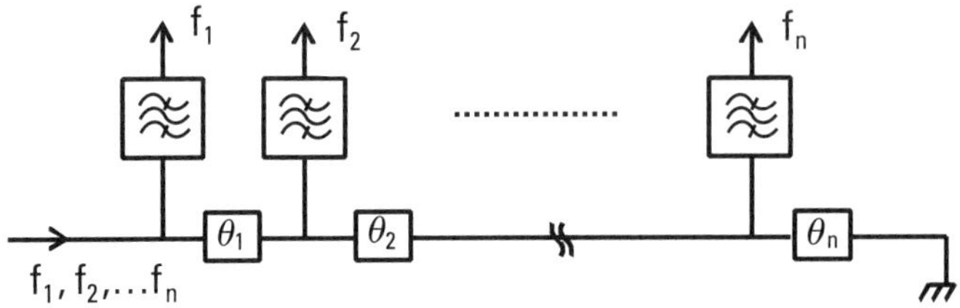

Figure 2.7 Schematic diagram of a manifold-coupled multiplexer configuration.

Manifold-coupled multiplexer. The manifold coupled multiplexer (Figure 2.7) does not use any directional properties. Consequently, it allows large interaction between channel filters. The design of this multiplexer requires modelling of all channel filters simultaneously in order to incorporate the effects of interchannel interactions, making its design very complex and inflexible to any change in frequency plan. Any change in frequency plan requires a redesign of this kind of multiplexer. The manifold-coupled multiplexer has the advantage of being the most compact design, with optimum erformance, both in terms of absolute insertion loss and amplitude and group delay response [13]. These characteristics make manifold-coupled multiplexers the best choice for realization of a high-power OMUX. Any number of channel filters with variable bandwidths can be combined using the manifold multiplexing approach, provided that the channel filters are spurious-free in the operating frequency band of the entire multiplexer, and have appropriate margins on either side.

2.2.1.2 IMUX/OMUX Filter Requirements

The channel performance is largely determined by the IM narrowband channel filters (typically between 0.2%–2.0% fractional bandwidth [10]). Other equipment tends to be broadband in nature. The wideband equipment contributes a small fraction to transmission amplitude and phase deviations (e.g., up to 5% [6]). Consequently, the stringent specification requirements are imposed mostly to the channel filters. The multiplexer channel filter requirements are also strongly dependent on the multiplexer realization architecture. The most commonly used approach for IMUX is the circulator coupled multiplexing (channel-dropping technique). With regard to the OMUX, the approach adopted depends on the frequency and the specification requirements of each payload. The multiplexers most commonly used are: the hybrid-coupled multiplexer, the circulator coupled multiplexer, the directional filter multiplexer, and the manifold-coupled multiplexer.

IMUX channel filters. The design drive is the efficient channelization of the signal into its various RF channels with a minimum loss of bandwidth. At the same time, enough isolation must be provided in order to control the interference from other channels. A minimum band guard is required that defines the overall usable bandwidth. Hence, very stringent out-of-band rejection requirements for the channel filters are imposed. Very high selectivity is most critical for preventing adjacent channel interference and multipath effects [6]. It is usually preferred that the bulk of the channel selectivity as the channel passes through the transponder is done by the input de-multiplexer (IMUX) filters to keep the output multiplexer (OMUX) filters as simple and low-loss as possible [13].

For the input demultiplexer (IMUX), the absolute value of the insertion loss is not a significant constraint (as long as it is within a few dBs) provided tha it is flat, (i.e., without significant variations in the passband). This is due to the fact that the noise figure is almost entirely controlled by the front-end receiver section and the following preamplifiers make up for the necessary gain in order to provide the desired input power to the high power amplifiers (HPAs). On the other hand, the distortion produced by the IMUX channel filters must be minimized, and therefore, flat insertion loss in the passband is critical. In traditional filter synthesis approach, this would in turn mean that the narrowband channel filters are high-Q filters since low Q would typically result into rounding of the pass-band edges. Alternatively, techniques to produce flat passbands from finite Q resonators can be employed [14]. Such stringent requirements reflect to the fact that the demultiplexer commonly carries the burden of providing most of the isolation between channels and suppressing interchannel interference. This is due to the fact that demultiplexers are at the low-power section of the transponders, and the system designers can afford putting the most stringent specifications at these components. Subsequently, the signal amplitude and phase distortion at the IMUX level is greater than the OMUX.

OMUX channel filter. At the high-power side of the payload, the individual channels are amplified and then recombined or multiplexed at the output multiplexer (OMUX). The channels for downlink must be efficiently combined into one high-power composite output and transferred to the input port of an antenna feed system. Now, the lowest possible loss is a priority, with a certain amount of selectivity to suppress out-of-band spectral regrowth caused by nonlinearities in the power amplifiers [7]. Both the OMUX channel as well as the output filters are required to comply with requirements for low PIM and high-power handling (multipaction).

Table 2.2 gives typical electrical specification requirements or input demultiplexers and output multiplexers including center frequency, absolute, and fractional bandwidth, insertion loss, insertion and group delay variation, out-of-band rejection levels, and power handling, as reproduced from Kunes [2] for a typical Ku-band system. This clearly describes the difference in the specifications for the high-power output multiplexer and the low-power high isolaton input demultiplexer. Table 2.3 shows typical channel filter specifications of a C-band OMUX (reproduced from [15]).

Finally, in designing a microwave filter for space, it is important to allow adequate margins for temperature variations. Temperature variations in a harsh environment where a spacecraft would move in and out of Earth's shadow would have significant impact in the electrical performance of very narrowband channel filters. The main effect is a shift of center frequency driven by thermal expansion, which (for aluminium filters) may be as much as 1 part in 1,000 over a 50°C temperature range. For Invar filters, the corresponding variation is about 1 in 10,000 [3]. Alternatively, the filter design may incorporate temperature compensation based on the use of two materials with different coefficients of expansion. Many filters are now made of

Table 2.2
Typical IMUX and OMUX Ku-Ba

Parameter	IMUX Specification	OMUX Specification
Operating frequency	14 GHz	12 GHz
Channel bandwidth	54 MHz	54 MHz
Fractional bandwidth	0.39%	0.45%
Insertion loss @ f0	11 dB	0.45 dB
IL variation within pass-band	<0.745 dB	<1.1 dB
Group delay variation within pass-band	<15 ns	<45 ns
Rejection at ±36 MHz of f_0	>35 dB	> 25 dB
Power handling	—	120 W/channel

Table 2.3
Typical Channer Filter Specifications of C-Band OMUX

Parameter	OMUX Specification
Operating frequency	3.7–4.2 GHz
Channel bandwidth	36 MHz (12 Channels)
Fractional bandwidth	0.95% (3.8 GHz)
Insertion Loss @ f0	≤ 0.5 dB
IL variation within pass-band	< 1.5 dB
Group delay variation within pass-band	< 65 ns
Rejection at ±25 MHz of f0	≥ 25/20* dB
Rejection at ±40 MHz of f0	≥ 40/35* dB
Power handling (average)	63 W/channel
PIM	–175 dBc
Operating temperature	–10°C to +65°C

*For noncontiguous edge of end channels

carbon-fiber-based materials [2]. These have both low mass and low thermal expansion, but their use in some applications is limited by difficulties in manufacturing.

2.2.2 Summary of Filters Encountered in a Communication Payload

In summary, the main filters typically employed within a communication payload are:

- Duplexer;
- Input filter;
- Demultiplexer—channel filters (IMUX);
- Multiplexer—channel filters (OMUX);
- Output filter;
- IF bandpass, low-pass, high-pass, and band-stop filters.

Duplexer. A duplexer is an essential component in a frequency division duplexing system (FDD), and is often encountered in communications satellite systems. It is the device that allows the system to utilize a common antenna to receive and transmit simultaneously by using a different frequency spectrum for the receive and transmit functions that are typically separated by a guard band. The duplexer is a three-port device, and consists of two filters with a common input (output) port and separate output (input) ports for each of the two filters. The guard band—in combination with the duplexer—ensures that there is enough isolation between transmit and receive signals. The main requirement on each filter is that it must provide adequate rejection of power levels at the other filter's passband frequency. For the receive filter, the rejection must be appreciable in order to reduce leakage of transmit power to the input mixer below a desired threshold. The transmit filter must prevent any significant wastage of receive energy in the transmitter, and must also adequately suppress unwanted transmitter modulation products, harmonics, and noise. The duplexer typically consists of two bandpass moderate bandwidth filters. For the receive section, the main requirement is low in-band loss to account for the low levels of the received signal coupled with adequate out-of-band attenuation. In the transmit section, low-loss and high levels of rejection at the receive band are required along with high-power handling capability.

Input filter. The input filter is a fairly wideband bandpass roofing filter that allows all the uplink frequencies to go through but eliminates out-of-band noise. The main requirements of the input filter include low in-band loss with adequate attenuation out-of-band. Low in-band loss is required because the received signal levels are very low. Typically, the input bandpass filter is followed by a low-pass filter.

Demultiplexer (IMUX). In the satellite transponder, the uplinked channel groups from a common antenna need to be separated (demultiplexed) in order to allow separate routing and processing before power amplification and downlink to (possibly) different zones on Earth. This is achieved by the input demultiplexer or IMUX. Each subfrequency band is processed by a single transponder. The channel bandpass filters have main requirements of rapid roll-off. Because it is critical to provide very low distortion to the amplitude and phase of the signals, a very flat in-band amplitude and group delay response is also required.

Multiplexer OMUX. The (OMUX) channel filters are bandpass, narrowband filters with main requirements of rapid roll-off together with low-loss and a sufficiently flat in-band amplitude and group delay response. Furthermore, high power handling and low PIM is needed.

Output filters. The output (harmonic rejection) filter are low-pass and/or bandpass filters with main requirements of low in-band loss with adequate attenuation at specified higher frequencies (second and third harmonics) and high power handling capability. They are typically employed within either of the following two configurations: as an individual filter per channel prior to the output multiplexer or as a filter for the combined output.

IF/various. In addition to what has been mentioned so far for the RF signal, intermediate-frequency amplification takes place at the lower microwave frequencies and, consequently, similar considerations apply to these filters. When compared with RF filters, the large fractional bandwidth at IF relaxes the performance requirements. Filters are also required in order to fulfil a variety of less exacting functions: bandpass filters that clean up the output spectrum of amplifiers; high-pass filters that provide high rejection at the fundamental frequency of a harmonic multiplier; and low-pass and band-rejection filters that are often used to supplement the rejection of a band-pass filter in a region where it is inadequate.

2.3 Satellite Filter Technology

Satellite filter technology has evolved substantially over the past decades [2, 7, 12]. The main challenge has always been the development of input and output channel filters. This is because the remaining microwave filters incorporated in the satellite payload pose minimum design challenges when compared to the channel filters. The advances in this area have been a technology-enabling factor for satellite communications [1]. The filter implementation technologies are broad and reflect the broad satellite application space and the wide and diverse frequency spectrum. This section gives an overview of the main technologies adopted by the satellite industry over the last forty years. Coaxial cavity filters and waveguide filters are commonly encountered technologies in space [10]. Important advances in filter technology for satellite filters include dual-mode waveguide filters [16] and dielectric resonator filters [7, 17, 18].

2.3.1 Coaxial Cavity Filters

Coaxial cavity filters are widely used in many satellite applications in the L- and S-bands. This class of filters represents a cavity implementation of narrowband combline filters [19]. Combline filters are coupled TEM-mode transmission lines where adjacent lines/resonators have the same short- and open-circuit reference plane [19–21]. A combline resonator comprises of a coaxial transmission line approximately equal to quarter-wavelength. Conventionally, it consists of a metallic post that is coaxially surrounded by a metallic enclosure/cavity. When partitions are employed between adjacent resonators, the combline filter evolves into a coaxial filter structure [21]. In order to narrow down the bandwidth, miniaturize the dimensions, and improve the filter selectivity, coupling slots/decoupling partitions can be utilized in a coaxial filter [20]. In particular, decoupling slots can be used to control the coupling between resonators that can be either magnetic or electric [20].

Coaxial cavity resonator technology can offer unloaded Q factor in the range of 2,000–5,000 for frequencies in, for example, the L-band. Since coaxial resonators are quarter-wavelength resonators (short-circuited at one end and open-ended at the opposite side), their first harmonic resonance occurs at approximately three times their fundamental frequency. This offers a wide spurious-free performance for this technology that can be a significant advantage in multiplexer design [13]. By capacitive loading the open end of the resonators, the spurious-free window can be further extended with values up to six times the fundamental frequency reported [22]. An additional key benefit of coaxial cavity filters is their very wide tuning range, which can be achieved by means of tuning screws. Transmission zeros at finite frequencies typically are realized by means of cross-couplings between nonadjacent resonators, and thus, they are usually employed in a folded structural configuration. Technological improvements in this area have also been driven by cellular radio developments [23], with the exception of advances relating to improved power handing capability [24].

Materials with stable thermal expansion coefficients (CTE)—such as Invar/Kovar—are commonly employed in order to avoid electrical performance instability with respect to temperature variations. Alternatively, temperature compensation countermeasures for coaxial cavity filters involve the combination of materials with different CTE for different parts of the metallic structure, (e.g. aluminum for cavity housing and Invar for resonator posts). Opposite thermal expansion coefficients can be exploited for compensation of the overall temperature dependence of the filters [25–28]. This technique is also widely used for helical resonator filters for space applications.

With respect to PIM product performance, some general guidelines are given by the telecommunications industry. These guidelines [2, 6, 8] apply to the design and manufacturing of coaxial cavity filters. In particular, tuning screws commonly utilized for post-fabrication tuning can significantly affect the PIM generation [29]. This is due to the contact interface between the cavity body and the screw that is commonly made of different material and is subject to worsening with the tuning process [29]. An important factor that affects PIM generation can be encountered in the excitation of coaxial cavity filters. The capacitive probe excitation technique (a metallic probe extended within the cavity towards the open end of the first coaxial resonator), where no galvanic contacts are required, is shown to be beneficial as compared to the inductive tap excitation (a metallic probe is extended within the cavity towards the open-end of the first coaxial resonator), where soldering is required. The excitation techniques are reviewed with specifics to helical resonators in Chapter 5, Section 5.2. Finally, it is commonly preferred that the cavity enclosure of the filters is made out of a single piece of metal. This can improve on PIM generation because it uses less metal to-metal contacts (even same metal materials) at the areas of high RF currents (the bottom surface of the cavity where the coaxial posts are connected to the cavity enclosure). The high power handling capabilities of coaxial cavity filters are substantially limited as compared to waveguide filter technology as a consequence of the miniaturization factor [24]. Techniques such as large-gap [6] (where larger physical distances are imposed in high field intensity regions), and/or dielectric filling techniques, (where low-loss high-power handling dielectric materials are used to mitigate the multipactor)[24, 30] are often applied.

2.3.2 Waveguide Technology

Early systems (C-band [6]) employed rectangular waveguide filters in single-mode operation, where a physical cavity supports a single resonant mode. Advantages of waveguide technology include the inherent high-power handling capabilities of waveguide structures, particularly at the lower part of the spectrum. Furthermore, the waveguide realization of the technology has allowed easy integration to the waveguide manifold, whict has been adopted as the most efficient multiplexer configuration [13, 31].

More recently—and as the satellite communications industry has gradually shifted to higher frequencies, in particular the Ku band (14/12 GHz) and above (e.g., Ka-band, 20/30 GHz)—the TE_{103}/TE_{113} rectangular/cylindrical cavity filters have become the industry's standard [2]. These filters have three times half-wavelength cavity lengths. A comprehensive review of this technology is reported in [31]. The adoption of dual-mode filters [16, 32, 33], where two orthogonal degenerate modes (e.g., TE_{111} in circular waveguide) are excited in the same physical cavity (half-wavelength long)—allowed for the reduction of the number of physical cavities by a factor of two with respect to the single-mode filters. The two modes in each cavity are coupled via a tuning screw or another type of discontinuity. Size reduction was one of the main benefits of this technology. This structure also allows the realization of transfer functions with transmission zeros because certain non-adjacent resonator couplings are inherent to the structure.

In order to ensure stable electrical performance in very narrowband waveguide channel filters with temperature variations, materials with very low CTE (such as Invar) are again required to build the waveguide components. Invar is a poor electrical conductor, and so all parts are typically silver-plated, usually with a good few skin depths to minimize losses. Invar is also not a lightweight material. Very thin (e.g., 0.5 mm) cavity walls are used to save mass because Invar is considered heavy. Aluminum is a lightweight material with improved electrical conductivity, but its CTE is many times that of Invar.

2.3.3 Dielectric Resonator Technology

High permittivity dielectric materials can offer reductions in size and weight compared to conventional waveguide filters. In 1982, a dual-mode axially mounted dielectric resonator loaded cavity filter was introduced [17], and its performance nearly matched the performance of conventional dual-mode waveguide filters. The structure layout is similar to dual mode waveguide filters with the exception of loading the empty cavities with dielectric pucks, thus reducing their physical size. This has marked the potential use of dielectric loaded multiplexers for space application; this technology is currently being successfully used on a number of satellites [7]. Subsequently, a planar dual-mode structure to realize dielectric loaded filters was invented [18]. This planar mounted dual-mode dielectric resonator has good mechanical and thermal stability, and is extremely flexible in the arrangement of cavities. This structure has found uses as the baseline design for the input multiplexer of C-band satellites [6].

2.3.4 Space Qualification for Filter Units

A range of electrical and mechanical performance tests are required in order for satellite payload equipment to be space qualified. For reference, Table 2.4 gives the electrical parameters monitored at these tests. Table 2.5 provides a list of the tests and a basic description of each test and its purpose.

Table 2.4
Electrical Parameters to be Measured During the Space Qualification Tests (at Hot, Cold, and Mid-Temp in Thermal Vacuum at Final Mid-Temp)

Parameter
Insertion loss (IL) versus frequency
Insertion loss (IL) slope versus frequency
Group delay versus frequency
Group delay (GD) stability
Selectivity
Data summary

Table 2.5
Space Qualification Tests on Microwave Equipment Onboard Satellite Payload

Test	**Description**
Initial Performance Test (IPT)	This test case provides the baseline data to compare with subsequent test results. The RF performance is measured to verify the compliance with the specification requirements.
Sine Survey	A modal survey test is conducted to identify the resonance frequencies of the DUT hard mounted on the test fixture through its nominal interface points. Resonance search is required at the beginning of the environmental test sequence and after the following steps: sine vibration test (qualification stage) and random vibration test (qualification stage and acceptance stage).
Sine Vibration	The DUT is subjected to low frequency excitations to demonstrate its ability to withstand the excitations of the launcher.
Health Check	To evaluate the health status of the DUT. To test after each vibration test (sine survey, sine vibrations); as unit level incoming test or during payload integration, whenever required.
Random Vibration	Simulates the random excitation produced by the launcher and the transmitted acoustic noise excitation. To demonstrate the ability to withstand the launch.
RF Post Vibration Test	To demonstrate that the DUT is not degraded after the preceding vibration tests. To compare with initial performance test (IPF).
Thermal Vacuum Cycling	Thermal vacuum cycling test, which simulates the extreme mission environment. To demonstrate the DUT's ability to survive in the space environment.
Final RF Test	An RF test at the end of the space qualification tests. Comparison between the IPF. The DUT must be compliant with the requirements.
Final Inspection	Optical and mechanical inspection of the DUT.

References

[1] Wray, L., G. Ommering van, and B. Kosinski, "Commercial Communications Satellites: A unique Business Model for Space Programs," in Proc. IEEE International Technology Management Conference, San Jose, CA, 2011.

[2] Kunes, M., "Microwave multiplexers for space applications," Electronics & Communications Engineering Journal, Vol. 10, No. 1, pp. 29–35, 1998.

[3] Fortescue, P., J. Stark, and G. Swinerd, Spacecraft Systems Engineering, Third Edition, Chichester, United Kingdom: Wiley, 2003, p. 704.

[4] European Space Agency (ESA), "Satellite frequency bands," Available: http://www.esa.int/Our_Activities/Telecommunications_Integrated_Applications/Satellite_frequency_bands. Last accessed May 2014.

[5] Pozar, D. M., *Microwave Engineering,* Third Edition, Hoboken, NJ: John Wiley & Sons, 2005.

[6] Kudsia, C., R. Cameron, and W.-C. Tang, "Innovations in Microwave Filters and Multiplexing Networks for Communication Satellite Systems," *IEEE Transactions on Microwave Theory and Techniques,* Vol. 40, No. 6, pp. 1133–1149, 1992.

[7] Cameron, R., C. Kudsia, and R. Mansour, *Microwave Filters for Communication Systems: Fundamentals Design and Applications,* New Jersey: Wiley-Interscience, 2007.

[8] Kunes, M., M. Gibson, G. Connor, J. Eskell, H. Ghuman, et al, "Low PIM Feedchain Design Techniques for Satellite Tranmit/Receive Antennas at L and Ku Band," in *Proceedings 19th European Microwave Conference,* London, 1989.

[9] Tang, W., and C. Kudsia, "Multipactor breakdown and passive intermodulation in microwave equipment for satellite applications," in *Proc. IEEE Military Communications,* Monterey, CA, 1990.

[10] Boria, V., and B. . Gimeno, "Waveguide filters for satellites," *IEEE Microwave Magazine,* Vol. 8, No. 5, pp. 60–70, 2007.

[11] Maral, G., M. Bousquet, and Z. Sun, *Satellite Communications Systems: Systems, Techniques, and Technologies,* Fifth Edition, Chichester, United Kingdom: Wiley, 2009.

[12] Craven, G., and C. Radcliffe, "Microwave filters for communication satellites," *Microwaves, Optics and Acoustics,* Vol. 2, No. 5, 1978.

[13] Cameron, R. J., and M. Yu, "Design of Manifold-Coupled Multiplexers," *IEEE Microwave Magazine,* Vol. 8, No. 5, pp. 46–59, October 2007.

[14] Guyette, A., I. Hunter, and R. D. Pollard, "The Design of Microwave Bandpass Filters Using Resonators With Nonuniform Q," *IEEE Transactions Microwave Theory and Techniques,* Vol. 54, No. 11, pp. 3914–3922, 2006.

[15] Singh, V., K. Parikh, S. Singh and R. Bavaria, "DROMUX for Satellite Communications," *IEEE Microwave Magazine,* Vol. 14, No. 6, pp. 104–118, September/October 2013.

[16] Atia, A., and E. Williams, "New types of waveguide bandpass filters for satellite transponders," *Comsat Technical Review*, Vol. 1, No. 1, Fall, 1971.

[17] Fiedziszko, S., "Dual mode dielectric resonator loaded cavity filter," *IEEE Transactions in Microwave Theory and Techniques*, Vol. 30, pp. 1311–1316, September 1992.

[18] Tang, W., et al., "Planar dual-mode cavity filters including dielectric resonators," March 24, 1987.

[19] Matthaei, G., L. Young, and E. Jones, *Microwave Filters, Impedance-Matching*, New York: McGraw-Hill, 1964.

[20] El Sabbagh, M., K. Zaki, H. Yao, and M. Yu, "Full-wave analysis of coupling between combline resonators and its application to combline filters with canonical configurations," *IEEE Transactions on Microwave Theory and Techniques*, Vol. 49, No. 12, December, 2001.

[21] Kurzrok, R., "Design of Comb-Line Band-Pass Filters (Correspondence)," *IEEE Transactions on Microwave Theory and Techniques*, Vol. 14, No. 7, pp. 351–353, July 1966.

[22] Yao, H., K. Zaki, A. Atia, and T. Dolan, "Improvement of spurious performance of combline filters," in *IEEE MTT-S International Microwave Symposium Digest*, Denver, CO, June 1997.

[23] Mansour, R., "Filter technologies for wireless base stations," *Microwave Magazine*, Vol. 5, No. 1, pp. 68–74, March 2004.

[24] Shamsaifar, K., T. Roddriguez, and J. Haas, "High-Power Combline Diplexer for Space," *IEEE Transactions on Microwave Theory and Techniques*, Vol. 61, No. 5, pp. 1850–1860, 2013.

[25] Wang, H., and K. Zaki, "Temperature compensation of combline resonators and filters," in *IEEE MTT-S International Microwave Symposium Digest*, Vol. 3, Anaheim, CA, June 1999.

[26] Yao, H.-W., and A. E. Atia, "Temperature Characteristics of Combline Resonators and Filters," in *Proc. IEEE International Microwave Symposium Digest*, Phoenix, AZ, 2001.

[27] Miiller, J., and M. Hoft, "Temperature Compensation of Resonators Using Different Materials and Suitable Dimensions," in *Microwave Conference, 2005 European*, Vol. 1, Paris, October 2005.

[28] Hoft, M., S. Burger, T. Magath, and O. Bartz, "Compact combline filter with improved cross coupling assembly and temperature compensation," in *Microwave Conference, 2006*, APMC 2006, Yokohama, Japan, Dec. 2006.

[29] Macchiarella, G., G. Stracca, and L. Miglioli, "Experimental Study of Passive Intermodulation in Coaxial Cavities for Cellular Base Stations Duplexers," *Proc.34th European Microwave Conference*, Amsterdam, 2004.

[30] Doumanis, E., G. Goussetis, W. Steffe, D. Maiarelli, and S. Kosmopoulos, "Helical Resonator Filters With Improved Power Handling Capabilities for Space Applications," *IEEE Microwave Components Letters*, Vol. 20, No. 11, pp. 598–600, 2010.

[31] Kudsia, C., J. Dorey, R. Heierli, K. Ainsworth, and G. Lo, "A New Type of Low Loss 14 GHz High Power Combining Network for Satellite Earth Terminals," in *Proc. 9th European Microwave Conference*, Brighton, United Kingdom, September 1979.

[32] Williams, A. E., and E. Atia, "Narrow-bandpass waveguide filters," *IEEE Transactions on Microwave Theory and Techniques*, Vol. 20, pp. 258–265, Apr. 1972.

[33] Williams, A. E., A. E. Atia, and R. Newcom, "Synthesis of dual-mode filters," *IEEE Transactions on Circuits and Systems*, Vol. 21, pp. 649–655, 1974.

3

Filter Synthesis Techniques

3.1 Introduction

At low frequencies, where the wavelength $\lambda = c/v$ is much longer than the circuit's physical dimensions, passive and analogue electrical filters are typically realized in lumped element technology (i.e., employing capacitors and inductors). These elements have simple frequency characteristics, and lumped element networks can be designed using circuit theory. At RF and microwave filters for space, the availability of lumped components can be restricted due one or more of the following reasons:

- *The distributed nature of the components.* As the wavelength shrinks at higher frequencies, it is no longer possible to manufacture the components that deliver the design impedance (both in terms of resistance and reactance) in a space confined to a small fraction of the wavelength.
- *The increased parasitics of packaged components.* Even if it is possible to fabricate lumped components, often their packaging and interconnections introduce unavoidable parasitic reactances that limit the use of lumped and/or packaged components. Likewise, resistive losses can become an increasing concern at higher frequencies.
- *The power handling requirements.* Lumped components typically have limited power handling capability. Circuits and filters required to withhold increased power levels comprise distributed elements that

reduce power densities by distributed given power levels to larger volumes.

For the aforementioned reasons, passive RF/microwave filters are commonly realized in distributed technologies. A distributed circuit element has one or more dimensions that are comparable with the wavelength. The frequency behavior of the elements is more complex, and the lumped element approach is no longer valid. Filters are, in this case, realized as distributed structures with carefully designed geometries. For distributed microwave filters, it is desirable that the synthesis would lead to a determination of physical dimensions of the device. A general and complete synthesis procedure for distributed filters is virtually impossible. However, a number of useful techniques has been developed for the design of microwave filters [1–5].

This chapter presents some basic underlying theory and practical techniques relating to the synthesis of distributed filters. The aim is not to provide an extensive theoretical background of this topic or advanced design techniques; for this purpose, the reader is encouraged to refer to extensive available literature [1, 4–6]. Instead, the aim here is to make available the basic theory and concepts that will enable the RF engineer to embark on practical design of microwave filters. Most of the material focuses on direct-coupled bandpass filters, although the concepts presented are the basis for more advanced filter topologies. Basic filter theory is reviewed with the emphasis on the analysis and applications of narrowband bandpass filters. Some common classical filter ladder network prototypes are discussed. Two filter network synthesis techniques are outlined, namely the circuit approach based on impedance/admittance inverters and the coupling matrix approach.

3.2 Synthesis of Prototype Response

Electrical filters, according to their filtering properties, are commonly distinguished in four types, namely: lowpass, highpass, bandpass, and bandstop. Figure 3.1 shows the ideal behavior of each class of filters. Other filter responses—such as all-pass filters that can be employed for equalization—are not discussed here. Ideal filter responses—such as those depicted in Figure 3.1—are not feasible in practice. It is therefore common to approximate the ideal transfer functions by the ratio of polynomials. Electrical filters can then be categorized with respect to the types of polynomials that are used to approximate the ideal transfer function. Although an infinite number of transfer functions can approximate the ideal filter behavior, common classes of filter responses are Chebyshev and Butterworth filters.

Most of the modern synthesis techniques are based on the insertion-loss method [1, 4–7] . According to this method, the filter response is initially defined as a function of frequency by means of its insertion loss. A lumped element network that gives such specified power loss function is then synthesized. The circuit model of the prototype filter provides the link to the physical realization of the filter networks. Based on the observation that all other classes of filters shown in Figure 3.1 can be obtained from the lowpass response upon a suitable frequency transformation, it is common practice in filter synthesis to synthesize the frequency response and lumped element network initially in terms of a lowpass prototype. Subsequently, the prototype of a bandpass filter can be obtained upon application of a suitable frequency transformation. This section summarizes the basic theoretical steps in the synthesis of bandpass prototype networks.

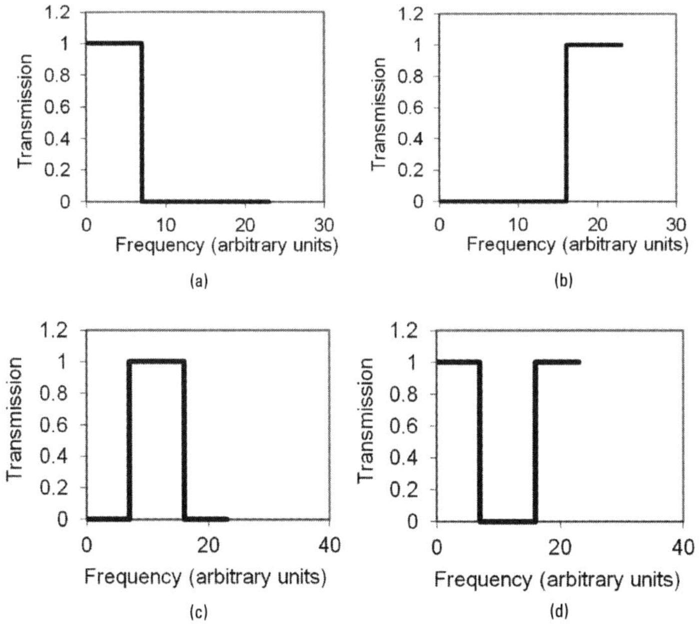

Figure 3.1 Ideal filter responses. (a) Ideal lowpass response, (b) ideal highpass response, (c) ideal bandpass response, and (d) ideal bandstop response.

3.2.1 Characteristics of Filter Response

The power loss ratio P_{LR} of a filter is defined as the ratio of the power available from the source divided by the power delivered to the load.

$$P_{LR} = \frac{\textit{Power available from the source}}{\textit{Power delivered to the load}} = \frac{P_{max}}{P_t} \tag{3.1}$$

Using $|\rho(\omega)|$ for the magnitude of the reflection coefficient as a function of the angular frequency, ω, then

$$P_{LR} = \frac{1}{1-|\rho(\omega)|^2} = \frac{1}{1-|S_{11}(\omega)|^2} = \frac{1}{|S_{21}(\omega)|^2} \tag{3.2}$$

where $S_{11}(\omega)$ and $S_{21}(\omega)$ follow the usual definition of the scattering parameters. It should be noted at this point that the and parameters are identical to the reflection and transmission coefficients, ρ, τ, so that

$$\rho = S_{11} \qquad \text{and} \qquad \tau = S_{21} \tag{3.3}$$

It is further common in the context of filters to refer to the reflection and transmission properties in terms of the transmission and reflection loss, often also termed as insertion loss (IL) and reflection loss (RL). The latter represents the reflection and transmission coefficient in the decibel scale according to:

$$IL = 10\log_{10}\frac{1}{|\tau(j\omega)|^2} = -20\log_{10}(|S_{21}|) \tag{3.4}$$

$$RL = 10\log_{10}\frac{1}{|\rho(j\omega)|^2} = -20\log_{10}(|S_{11}|) \tag{3.5}$$

The reflection coefficient $|\rho(\omega)|^2$ is an even function of ω ([8], Section 4.1). $|\rho(\omega)|^2$ can therefore be expressed as a polynomial in ω^2, so that:

$$|\rho(\omega)|^2 = \frac{M(\omega^2)}{M(\omega^2)+N(\omega^2)} \tag{3.6}$$

where M and N are real polynomials in ω^2. By substituting (3.6) in (3.2), the power loss ratio can be expressed as:

$$P_{LR} = 1 + \frac{M(\omega^2)}{N(\omega^2)} \tag{3.7}$$

Physically realizable filters have response that can take the form of (3.7). It is convenient to define a characteristic *function*, $K(s)$, so that P_{LR} can be expressed as:

$$P_{LR}(s) = 1 + \left|K(s)\right|^2_{s=j\omega} \tag{3.8}$$

where we have also reverted to the complex frequency variable definition, $s(s = \sigma + j\omega)$ [4]. Because the power reflected from the filter represents the power available from the source P_{max} reduced by the power delivered to the load, P_t

$$P_r = P_{max} - P_t \tag{3.9}$$

it is evident that the characteristic function represents the ratio of reflected over transmitted power:

$$\left|K(s)\right|^2_{s=j\omega} = \frac{\textit{reflected power}}{\textit{transmitted power}} = \frac{P_r}{P_t} = \frac{\left|\rho(j\omega)\right|^2}{\left|\tau(j\omega)\right|^2} \tag{3.10}$$

To analyze the characteristic function $K(s)$ and the properties of the reflection and transmission coefficients, it is useful to consider the input impedance properties of realizable networks. For any physical electrical network comprising linear and time invariant R, L, and C elements, it can be shown that the input impedance is a positive real function ([4], Section 2.5). As such, the input impedance can be expressed in the form of a ratio of polynomials with real, rational and positive coefficients. Thus:

$$z(s) = \frac{n(s)}{d(s)} \tag{3.11}$$

Expressing the reflection coefficient in terms of the normalized input impedance of the prototype network Z_L according to [8]:

$$\rho = \frac{z_L - 1}{z_L + 1} \tag{3.12}$$

suggests that the reflection coefficient can take the form:

$$\rho(s) = \frac{n(s) - d(s)}{n(s) + d(s)} = \frac{F(s)}{E(s)} \tag{3.13}$$

where $F(s)$, $E(s)$ are polynomials. Because the impedance $z(s)$ is represented by a positive real number, the coefficients of $E(s)$ and $F(s)$ are also real. In order to obtain the form of the transmission coefficient, we consider that (for lossless passive networks) the reflection and transmission coefficients are related according to:

$$|\tau(s)|^2 = 1 - |\rho(s)|^2 \tag{3.14}$$

so that

$$|\tau(s)|^2 = 1 - \frac{|F(s)|^2}{|E(s)|^2} = \frac{|E(s)|^2 - |F(s)|^2}{|E(s)|^2} \tag{3.15}$$

Chapter 3 [4] shows that the nominator can be expressed as a square of polynomial, $P(s)$, defined by:

$$|P(s)|^2 = |E(s)|^2 - |F(s)|^2 \tag{3.16}$$

so that the transmission coefficient takes the form:

$$|\tau(s)| = \frac{|P(s)|}{|E(s)|} \tag{3.17}$$

The polynomials $F(s)$, $E(s)$, and $P(s)$ are known as characteristic polynomials. For physical lowpass prototype filter networks, we can write in summary:

$$\rho(s) = S_{11}(s) = \frac{F(s)}{E(s)} \quad \text{and} \quad \tau(s) = S_{21}(s) = \frac{P(s)}{E(s)} \tag{3.18}$$

In view of (3.10) it also holds that

$$K(s) = \frac{F(s)}{P(s)} \tag{3.19}$$

The properties of the characteristic polynomials define the filter characteristics. $E(s)$ is a Hurwitz polynomial with roots lying in the left half of the s-plane or on the imaginary axis (i.e., real(s) ≤ 0). These roots correspond to the singularities of the reflection and transmission coefficients, and represent the natural frequencies (frequencies where oscillations are sustained in the absence of an external source, also known as eigenfrequencies) of the filter. The fact that they lie in the left half of the s-plane (or in the limit on the imaginary axis) represents the passivity of the filter because any natural oscillations associated with frequencies in the right half s-plane would be of exponentially increasing magnitude with respect to time. Roots of $P(s)$ can exist in conjugate pairs along the imaginary axis, $s = j\omega$. These roots represent angular frequencies at which all power is reflected and are known as *transmission zeros* or *attenuation poles*. The roots of $F(s)$ occur at the origin (potentially multiple) and in conjugate pairs along the imaginary axis. They correspond to frequencies where all incoming power is transmitted and are sometimes termed as reflection zeros.

By combining (3.2) and (3.8), it can also be seen that the general expressions for the magnitude of reflection and transmission (or S_{11}, S_{12} respectively) of a filter are:

$$\left|S_{11}(j\omega)\right|^2 = \frac{\left|K(j\omega)\right|^2}{1+\left|K(j\omega)\right|^2} \quad \text{and} \quad \left|S_{21}(j\omega)\right|^2 = \frac{1}{1+\left|K(j\omega)\right|^2} \tag{3.20}$$

Beyond the distinction in terms of its lowpass, bandpass, bandstop, and highpass response, filters are also classified in terms of the characteristic function, $K(s)$, of the prototype response. Commonly used filters employing include Butterworth (maximally flat), Chebyshev (equiripple) filters, and elliptical filters. They are reviewed in the following sections in the context of lowpass filters.

3.2.2 Chebyshev Response

Chebyshev filters produce equal-ripple in the passband. They are defined by the characteristic function

$$|K(j\omega)|^2 = \varepsilon^2 T_n^2\left(\frac{\omega}{\omega_c}\right) \tag{3.21}$$

where n is the order of the filter, ω_c is the cut-off frequency, ε is the ripple coefficient, and T_n represents the Chebyshev polynomial of the first kind and nth order:

$$T_n(x) = cos\left[n \bullet arccos(x)\right] = cosh\left[n \bullet arccosh(x)\right] \tag{3.22}$$

Chebyshev filters have all transmission zeros located at infinity, while the poles lie along an ellipse in the left half s-plane. The main parameters that characterize the response of an ideal Chebyshev filter are the order, n, and the ripple coefficient, ε, that (in the case of a lossless filter) determines the maximum insertion loss in the passband. It is therefore common to express the ripple coefficient in terms of the ripple level that describes the maximum in-band insertion loss, I_L, for the lossless case:

$$IL = 10log_{10}\left(1+\varepsilon^2\right) \tag{3.23}$$

Figure 3.2(a, b) plots the Chebyshev polynomials of the third- and fourth-order, whereas Figure 3.2(c, d) plots the corresponding ideal Chebyshev filter responses for ripple level 1. Furthermore, the filter response has been plotted in both negative and positive frequencies (although for lowpass filters, the negative frequency range has limited physical meaning; it can have an effect upon the application of frequency mapping, such as the one required for deriving the bandpass prototype from the lowpass prototype).

Figure 3.3 shows a third-order filter of ripple level 0.5, 0.1, and 0.01 dB, respectively. The ripple level is usually determined by the maximum allowed insertion loss requirement given in the filter specifications, although it also affects the filter sharpness. Figure 3.4 shows the first three ideal Chebyshev filters of ripple 0.5 dB. As seen on this plot, the order, n, determines the number of ripples of the insertion loss in the passband and is crucial for the

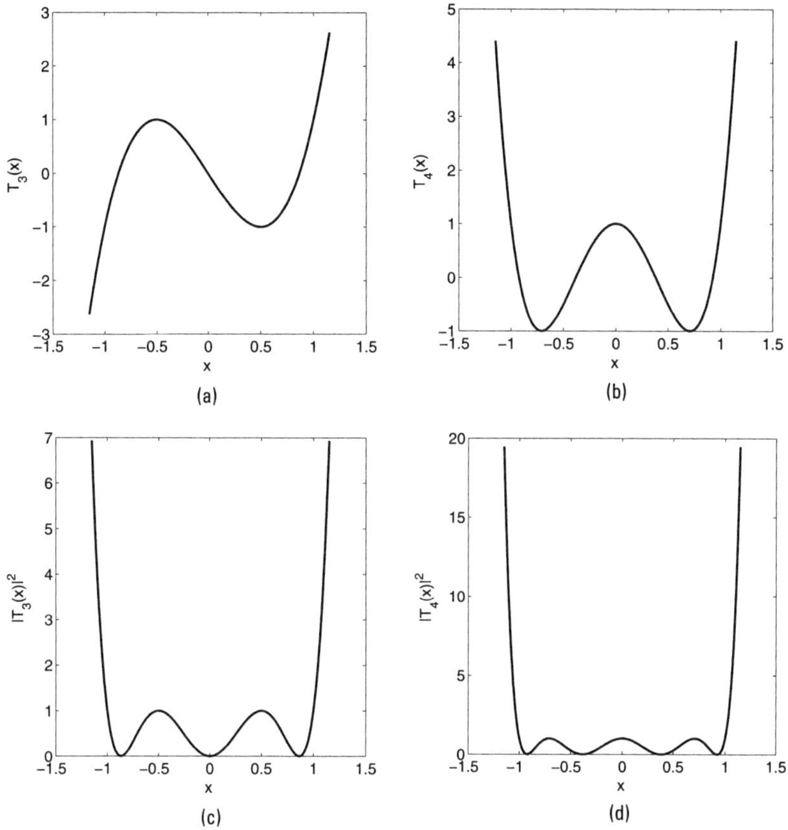

Figure 3.2 Chebyshev polynomials of the (a) third and (b) fourth order. The corresponding filter response for ripple 1 are shown in (c) and (d).

filter's skirt selectivity. It is commonly determined by the out-of-band attenuation requirements and the chosen ripple.

3.2.3 Butterworth Response

Butterworth filters produce maximally flat passbands, typically at the cost of reduced skirt selectivity when compared to Chebyshev filters. They are defined by the characteristic

$$\left|K(j\omega)\right|^2 = \omega^{2n} \tag{3.24}$$

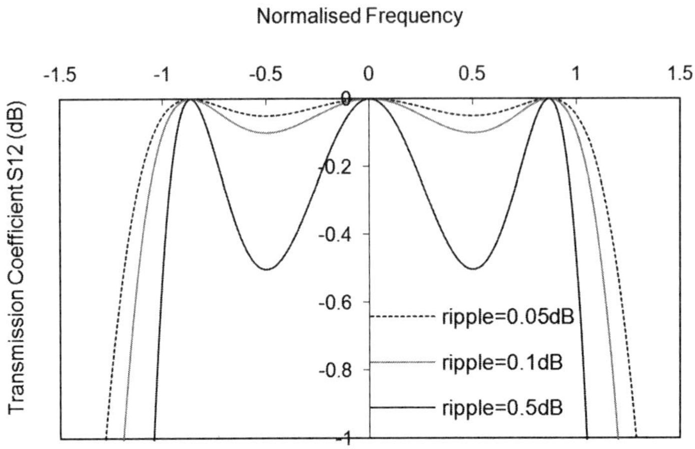

Figure 3.3 Chebyshev filter response of various ripple levels (third-order).

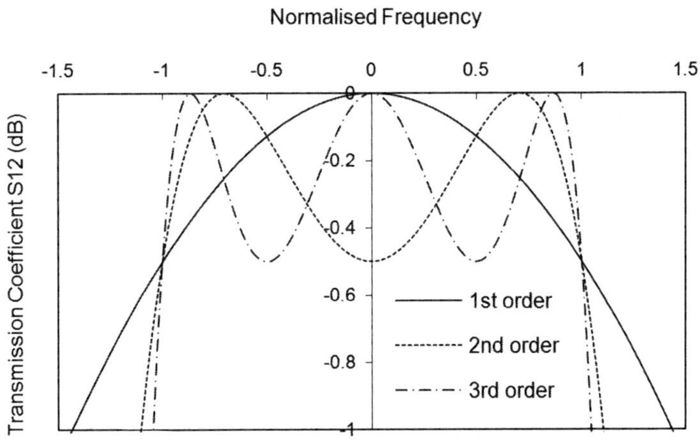

Figure 3.4 Chebyshev filter response of various order (ripple 0.5 dB).

where n is the order of the filter. Similar to Chebyshev filters, all transmission zeros lie at infinity; as a result, Chebyshev and Butterworth filters are often termed as *all-pole* filters. The poles of Butterworth filters lie along a unit circle in the left half s-plane at equally spaced angles. Higher-order filters produce steeper selectivity. The order of a Butterworth filter can be determined by the insertion loss at the cutoff frequency and a minimum insertion loss value at a frequency in the stopband. The power loss ratio and the insertion loss of Butterworth filters with insertion loss at the cutoff frequency at 2- and 3-dB level, respectively, are plotted in Figure 3.5 as a function of normalized frequency. For comparison, the response of a fourth-order lowpass Chebyshev prototype with ripple levels 2 and 3 dB, respectively, is superimposed.

3.2.4 Elliptic and Asymmetric Responses

Both Chebyshev and Butterworth responses present no transmission zeros at finite frequencies, but instead have all the transmission zeros located at infinity. These filters are therefore commonly termed as all-pole filters. All-pole responses can be implemented by ladder networks (discussed in Section 3.3). Modern communication systems often benefit from filter responses that introduce transmission zeros at finite frequencies, so that the skirt selectivity of the filter is improved and/or increased isolation from a known out-of-band interferer is achieved. *Elliptical responses* correspond to a class of filters that produce equal ripple both in the passband and in the stopband. Unlike all-pole filters, the implementation of elliptical filters requires resonant LC circuits in a series or shunt branch of the prototype network. Similar to Chebyshev and Butterworth responses, elliptical lowpass filters produce symmetrical response with respect to the origin.

For some applications, asymmetric frequency response can be preferable. Lowpass *asymmetric response* cannot be implemented using conventional RLC networks, although this may be possible for bandpass asymmetric response. Unlike elliptic and all-pole responses, the characteristic polynomials $P(s)$ and $F(s)$ (3.18) do not necessarily have conjugate roots, and their coefficients are no longer real (although $E(s)$ is still a Hurwtiz polynomial with roots in the left half s-plane in order to respect passivity).

The implementation of lowpass filter prototype networks with elliptical or asymmetrical responses is not possible with the traditional ladder networks but requires more advanced techniques (discussed next in Section 3.3). The description of these methods extends beyond the scope of this book. For more details about advanced filter synthesis techniques, the reader is referred to the relevant literature provided throughout this chapter. We will therefore elaborate no further on the mathematical description of the associated responses.

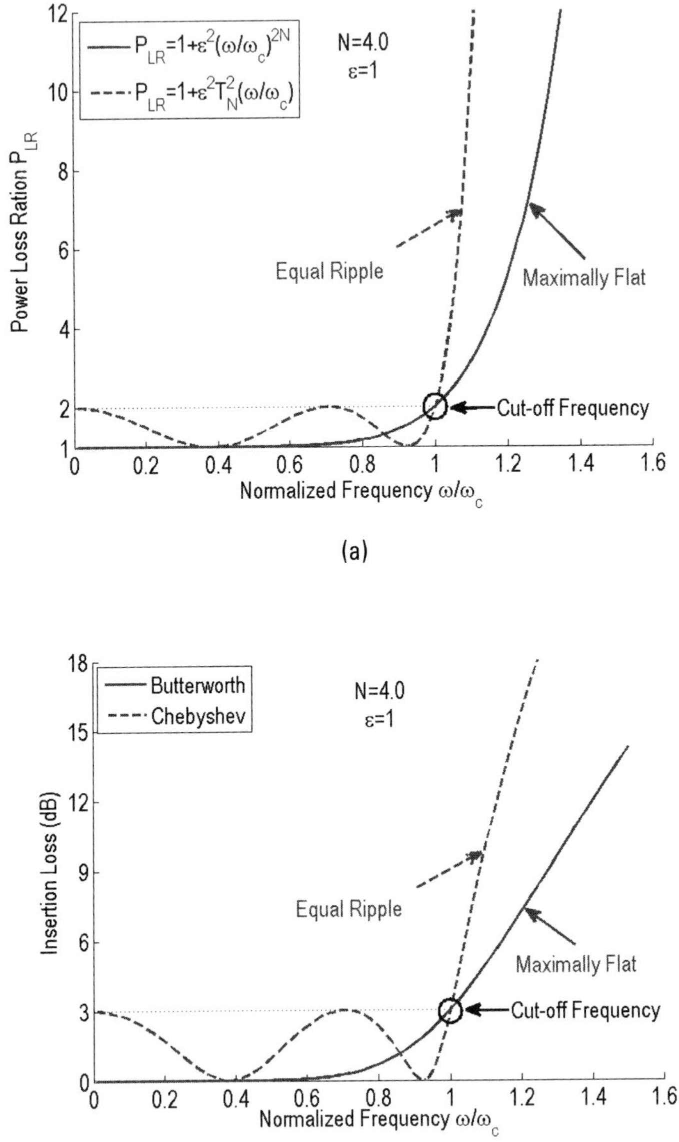

Figure 3.5 Power loss ratio and insertion loss for a fourth-order Butterworth and Chebyshev prototype filter.

3.3 Bandpass Filters Synthesis Using a Ladder Network

Once the characteristic function and polynomials have been determined, the next step in the design process is to translate these polynomials into a prototype electrical circuit from which a practical RF/microwave filter may be developed. The aim of this section is to present a practical step-by-step method for the synthesis of distributed all-pole bandpass filters. In particular, starting from an equivalent lowpass lumped element prototype (Section 3.3.1), the transformation to an equivalent circuit with only series or shunt elements by virtue of using immittance inverters is discussed in Section 3.3.2; the latter is easier to implement in practice using distributed structures of either series- or shunt-only elements. Subsequently, the frequency transformation that enables the conversion of a lowpass response into a bandpass response is presented in Section 3.3.3. The implementation of these circuits using distributed resonators (e.g., TEM lines or waveguides) is discussed in [6] and is not detailed here. Likewise, for the detailed synthesis of more advanced filter responses, the reader is referred to dedicated literature on filter synthesis [1, 4–6]. It is noted that the circuit prototypes developed in this section can directly be used for the extraction of the coupling matrix, as discussed in the Section 3.4.

3.3.1 Lowpass Lumped Element Filter Prototype

General forms of *n*th-order lowpass prototypes suitable for the realization of all-pole filters (e.g., Butterworth and Chebyshev responses) are the ladder networks shown in Figure 3.6. It is common to consider the prototype lowpass network normalized to source resistance of 1 Ω and a lowpass characteristic with its band-edge frequency at $\omega_c' = 1\ rad/\sec$. The values of the normalized network, g_n, indiscriminately represent inductance, capacitance, resistance, or conductance. They can be scaled to actual values that account for the actual source resistance (see also discussion in Section 3.3.2). Likewise, a lowpass-to-bandpass transformation is used to map the lowpass prototype network with unitary cutoff frequency to a bandpass prototype network that accounts for frequencies associated with the required bandpass response (see also Section 3.3.3). The two columns in Figure 3.6 represent shunt-series and series-shunt implementations; the two are dual to each other and will provide the same response for the same normalized element values, g_n. The two rows represent the even and odd orders of the lowpass prototype network, *n*. The use of lossless elements simplifies the synthesis process.

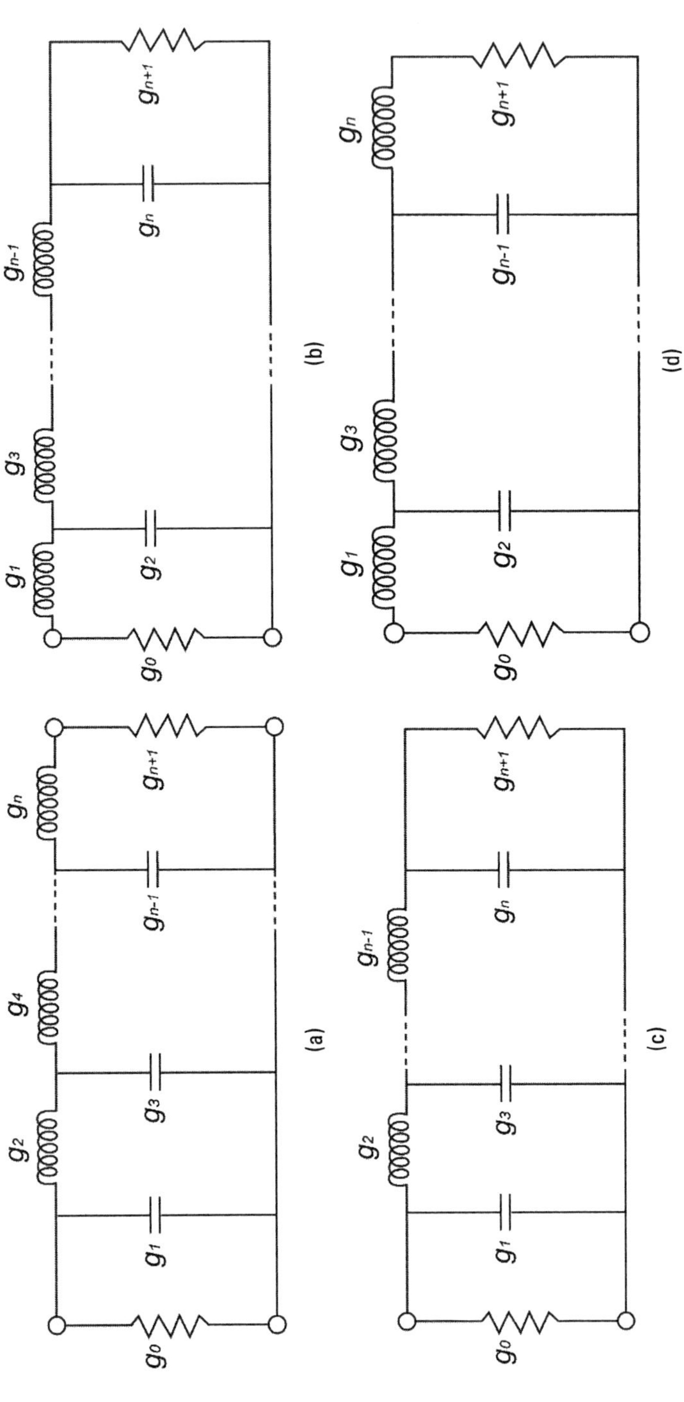

Figure 3.6 General form of lowpass prototype ladder filters: (a) shunt-series of even order n, (b) series-shunt of even order n, (c) shunt-series of odd order n, (d) series-shunt of even order n.

For Butterworth and Chebyshev lowpass filter responses, closed form expressions for the g_k values (as depicted in Figure 3.6) are available [1]. The normalized element values for the Chebyshev lowpass prototype are given in (3.25)

$$g_o = 1$$

$$g_1 = \frac{2\alpha_1}{\gamma}$$

$$g_k = \frac{4\alpha_{k-1}\alpha_k}{b_{k-1}g_{k-1}}, \text{ for } k = 2, 3, \ldots, n$$

$$g_{n+1} = \begin{cases} 1, & \text{for } n \text{ odd} \\ \coth^2\left(\frac{\beta}{4}\right), & \text{for } n \text{ even} \end{cases}$$

$$\alpha_k = \sin\frac{(2k-1)\pi}{2n}$$

$$\beta = ln\left[cot\left(\frac{L_{Ar}}{17.37}\right)\right]$$

$$\gamma = sinh\left(\frac{\beta}{2n}\right)$$

$$b_k = \gamma^2 + \sin^2\frac{k\pi}{n} \tag{3.25}$$

The normalized element values for the Butterworth lowpass prototype are given in (3.26) through (3.27):

$$g_o = 1$$

$$g_k = 2 \bullet sin\left[\frac{(2k-1)\pi}{2n}\right], \qquad \text{for } k = 1 \text{ to } n$$

$$g_{n+1} = 1 \tag{3.26}$$

It is noted that prototype networks used to implement other filter responses can be synthesized using network synthesis techniques, such as those described in [4, 6]. For generalized transfer functions when transmission zeros are not located at infinity, additional section elements are required (other than capacitor and inductor sections).

As discussed earlier, the lowpass prototype networks presented above are normalized to source resistance equal to 1Ω. In practice, the source resistance can be arbitrary, R_o. For this case, the actual values for the inductors and capacitors can be obtained from the normalized element values, g_k, and the source resistance, R_o, according to:

$$L = g_k \bullet R_o$$

$$C = \frac{g_k}{R_o}$$

$$R = g_{n+1} \bullet R_o \tag{3.27}$$

It is noted that with the exception of Chebyshev prototypes of the even order, the load resistance for the all-pole prototypes discussed here—which is obtained by (3.35)—is unity.

3.3.2 Immittance inverter Lowpass Lumped Prototype

A difficulty in the design of distributed filtering structures is that an arrangement consisting of alternate series and shunt elements is often difficult to realize. The concept of immittance (i.e., impedance or admittance) inverters has been introduced in filter synthesis [9] as a tool to transform the ladder networks depicted in Figure 3.6 (which, in general, include both series inductances and shunt capacitances) to equivalent networks that include either series inductors or shunt capacitors. The former is achieved with the use of impedance inverters (also commonly referred to as K-inverters), while the latter exploits admittance inverters (also known as J-inverters).

An impedance inverter is a theoretical, frequency-independent, symmetrical two-port network, described by its characteristic impedance, K. When terminated with impedance Z_0 at one port, the impedance seen at the other port, Z_i is given by

$$Z_i = \frac{K^2}{Z_o} \tag{3.28}$$

This is schematically depicted in Figure 3.7. From (3.28), it stems that a shunt capacitor with a K-inverter connected on either side looks like a series inductance from its exterior terminals. Hence, K-inverters allow for the replacing all shunt capacitances of the lumped element filter prototype of Figure 3.6, with series inductances connected between K-inverters upon an appropriate choice of the scaling factor. It can be proven [1] that an equivalent circuit to that shown on Figure 3.6 is the one shown on Figure 3.8, provided that

$$K_{0,1} = \sqrt{\frac{R_o L_1}{g_o g_1}}$$

$$K_{k,k+1} = \sqrt{\frac{L_k L_{k+1}}{g_k g_{k+1}}}$$

$$K_{n,n+1} = \sqrt{\frac{L_n R_L}{g_n g_{n+1}}} \tag{3.29}$$

Observation of (3.29) reveals that arbitrary inductance values can be chosen in the network topology of Figure 3.8, provided that the K-inverter values are selected appropriately. It is easy to show that the ABCD matrix of a K-inverter is of the form:

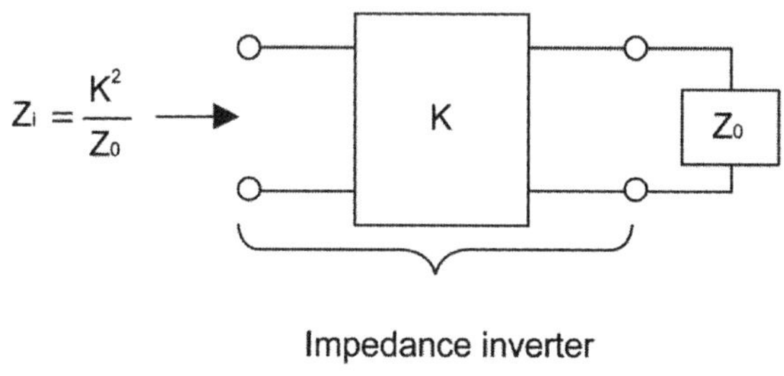

Figure 3.7 Schematic block diagram of K-inverter.

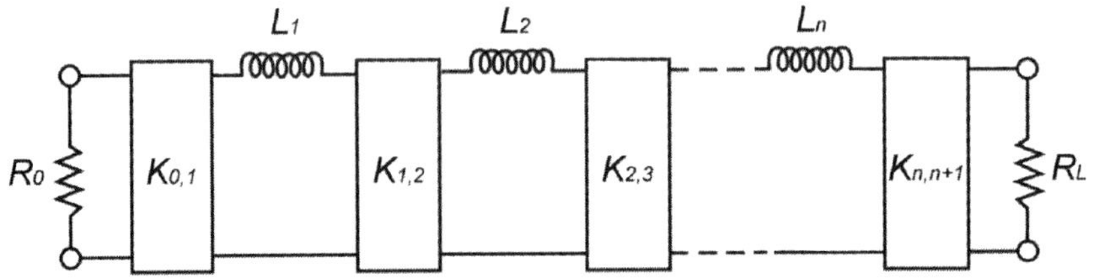

Figure 3.8 Modified prototype of Figure 3.6 using impedance inverters.

$$\begin{bmatrix} A & B \\ C & D \end{bmatrix} = \begin{bmatrix} 0 & \mp jK \\ \pm \frac{j}{K} & 0 \end{bmatrix} \tag{3.30}$$

It is noted that a quarter wavelength transmission line performs as a K-inverter at a single frequency point. Other implementations include T-networks with shunt positive and series negative inductors/capacitors [1].

The admittance inverter (commonly referred to as J-inverter) is the dual of a K-inverter. When an admittance, Y_o, is connected at one terminal of a J-inverter with characteristic admittance J, the admittance seen at the other port, Y_i is given by:

$$Y_i = \frac{J^2}{Y_o} \tag{3.31}$$

The use of admittance inverters allows transforming the lowpass prototype network of Figure 3.6 to that of Figure 3.9 provided the following values for the J-inverters are used:

$$J_{0,1} = \sqrt{\frac{Y_o C_1}{g_o g_1}}$$

$$J_{k,k+1} = \sqrt{\frac{C_k C_{k+1}}{g_k g_{k+1}}} \tag{3.32}$$

$$J_{n,n+1} = \sqrt{\frac{C_n Y_L}{g_n g_{n+1}}}$$

It is easy to show that the ABCD matrix of a K-inverter is of the form:

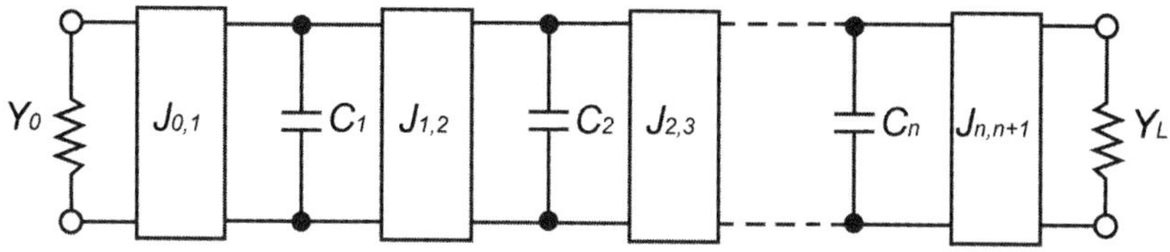

Figure 3.9 Modified prototype of Figure 3.6 using admittance inverters.

$$\begin{bmatrix} A & B \\ C & D \end{bmatrix} = \begin{bmatrix} 0 & \pm\dfrac{1}{jJ} \\ \mp jJ & 0 \end{bmatrix} \tag{3.33}$$

3.3.3 Frequency Transformation

As discussed in Section 3.3.1, the lowpass prototype presented so far is normalized to cutoff frequency at $\omega_c' = 1\ rad/\sec$. In order to scale this to the desired cutoff frequency, ω_c, the following transformation should be applied:

$$\omega' = \frac{\omega}{\omega_c} \tag{3.34}$$

where ω' is the normalized angular frequency and ω is the actual angular frequency. This transformation also applies to the reactances and susceptances of the prototype network. Considering also the normalization with respect to the source load, the final inductance and capacitance values of the lowpass prototype that produces the desired frequency response for the actual source impedance are given by:

$$L = \frac{g_k \bullet R_o}{\omega_c} \tag{3.35}$$

$$C = \frac{g_k}{\omega_c R_o}$$

Various frequency transformations are available for conversion of the lowpass prototype to highpass, bandstop, and bandpass, respectively. In the remaining paragraphs, we will consider the lowpass-to-bandpass transformation; for other cases the reader is referred to [5].

Bandpass response is obtained from the lowpass prototype by transforming the frequency scale in such a way that the center frequency ω_o of the desired bandpass corresponds to $\omega' = 0$ of the lowpass, and the band-edge frequencies ω_1 and ω_2 of the bandpass filter correspond to the band edge of the prototype = 1. This is achieved by the following frequency transformation:

$$\omega' = \frac{1}{FBW}\left(\frac{\omega}{\omega_o} - \frac{\omega_o}{\omega}\right) \tag{3.36}$$

where FBW and ω_o stand for the fractional bandwidth and geometric mean associated with the lower and higher cutoff frequencies:

$$FBW = \frac{\omega_2 - \omega_1}{\omega_o} \tag{3.37}$$

$$\omega_o = \sqrt{\omega_2 \bullet \omega_1}$$

This transformation can be applied to the lowpass prototype shown in Figure 3.8 in order to derive a bandpass prototype with series-only type of resonators. Because the impedance inverters (K-inverters) are frequency-independent elements, they are not affected by this frequency transformation. Instead, the series inductors of the lowpass prototype of Figure 3.8 now become series LC resonators, as shown in the bandpass K-inverter prototype of Figure 3.11. Considering the transformation of (3.27), the inductor and capacitor values of the bandpass circuit of Figure 3.11, L_{sk} and C_{sk} respective-

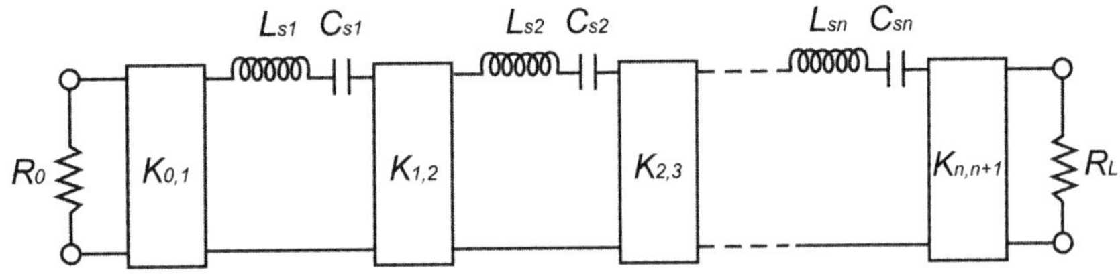

Figure 3.10 Lumped element bandpass prototype of the nth-order employing K-inverters.

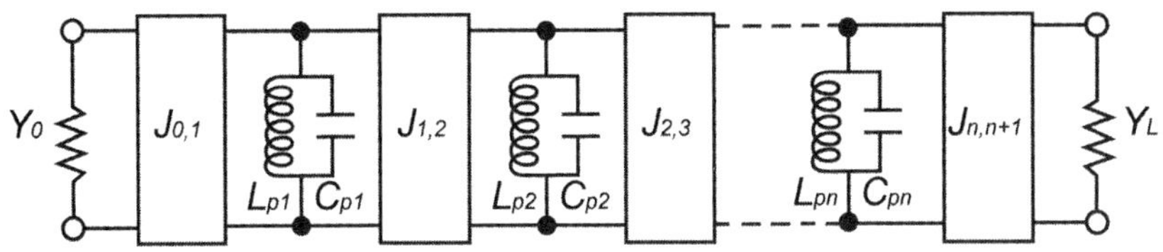

Figure 3.11 Lumped-element bandpass prototype of the *n*th-order employing *J*-inverters.

ly, will be related to the inductor values, L_k, of the lowpass circuit of Figure 3.8 according to

$$L_{sk} = \frac{L_k}{FBW \bullet \omega_o} \qquad C_{sk} = \frac{FBW}{L_k \bullet \omega_o} \tag{3.38}$$

Since the *K*-parameters of (3.29) remain unchanged, we can substitute for L_k according to (3.27) through (3.34) and thus obtain for the bandpass lumped element prototype with *K*-inverters of Figure 3.11:

$$K_{0,1} = \sqrt{\frac{FBW \bullet \omega_o R_o L_{s1}}{g_o g_1}} \qquad K_{k,k+1} = FBW \bullet \omega_o \sqrt{\frac{L_{sk} L_{sk+1}}{g_k g_{k+1}}} \qquad K_{n,n+1} = \sqrt{\frac{FBW \bullet \omega_o L_{sn} R_L}{g_n g_{n+1}}} \tag{3.39}$$

As with the *K*-inverter lowpass prototype network, arbitrary values can be chosen for the series inductors, Lsk. In this case, however, the capacitor values should then be such that the series LC resonators resonate at the filter's center frequency, ω_o:

$$C_{sk} = \frac{1}{\omega_o^2 L_{sk}} \tag{3.40}$$

The same transformation can be applied to the lowpass prototype network with J-inverters and shunt capacitors (Figure 3.9) in order to obtain a bandpass prototype with shunt-only type of resonators. The latter is shown in Figure 3.12.

Following a similar approach, it can readily be shown that (in this case) the characteristic admittances of the *J*-inverters are given by:

$$J_{0,1} = \sqrt{\frac{FBW \bullet \omega_o Y_o C_{p1}}{g_o g_1}}$$

$$J_{k,k+1} = FBW \bullet \omega_o \sqrt{\frac{C_{pk} C_{pk+1}}{g_k g_{k+1}}}$$

$$J_{n,n+1} = \sqrt{\frac{FBW \bullet \omega_o C_n Y_L}{g_n g_{n+1}}} \tag{3.41}$$

while the inductors values are such that the shunt LC resonators resonate at the filters' center frequency, ω_o:

$$L_{sk} = \frac{1}{\omega_o^2 C_{sk}} \tag{3.42}$$

It is noted that the lowpass-to-bandpass transformation can also be applied to the prototype network of Figure 3.6, yielding an equivalent lowpass

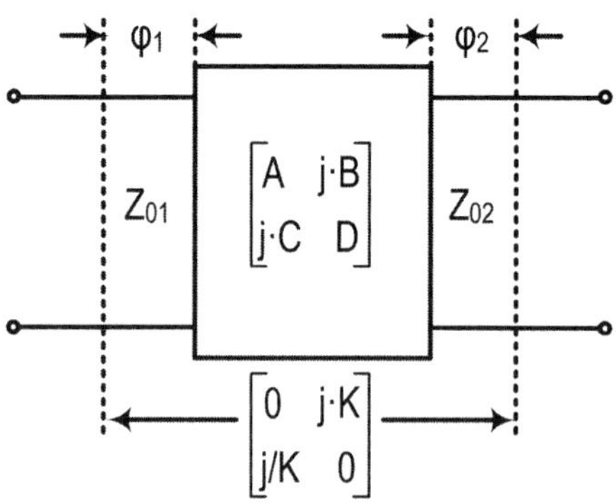

Figure 3.12 Convert lossless ABCD matrix into *K*-inverter.

lumped-element prototype that consists of alternating series and shunt LC resonators.

3.3.4 Impedance Inverters from a Discontinuity

A range of circuits that realize impedance inverters are available [1] that can be particularly useful in circuit synthesis of filters. More generally, it is described in [10] that any lossless reciprocal passive networks can be converted to an equivalent K-inverter. In particular, it is shown in [10] that it is always possible to choose reference planes at electrical distances φ_1 and φ_2 from the reference planes of the structure into an impedance inverter. For completeness, this proof is summarized below.

Assume the lossless passive reciprocal two-port network with transfer matrix ABCD as shown in Figure 3.12. The characteristic impedance of the two ports are Z_{01} and Z_{02}, respectively. The transfer matrix of the total two-port network defined between the two reference planes at electrical distances φ_1 and φ_2 is given by the product of the partial transferring matrices:

$$\begin{bmatrix} A' & jB' \\ jC' & A' \end{bmatrix} = \begin{bmatrix} \cos\varphi_1 & jZ_{01}\sin\varphi_1 \\ \dfrac{j\sin\varphi_1}{Z_{01}} & \cos\varphi_1 \end{bmatrix} \bullet \begin{bmatrix} A & jB \\ jC & D \end{bmatrix} \bullet \begin{bmatrix} \cos\varphi_2 & jZ_{02}\sin\varphi_2 \\ \dfrac{j\sin\varphi_2}{Z_{02}} & \cos\varphi_2 \end{bmatrix} \tag{3.43}$$

This 2-port is an impedance inverter if, by comparison (Figure 3.12), $A' = D' = 0$. Upon algebraic manipulation it turns out that for selection of the reference plane locations at electric distances ϕ_1 and ϕ_2 satisfying:

$$\tan 2\varphi_1 = \frac{2\left(\dfrac{BD}{Z_{02}} - ACZ_{02}\right)}{A^2\dfrac{Z_{02}}{Z_{01}} + \dfrac{B^2}{Z_{01}Z_{02}} - C^2 Z_{01}Z_{02} - D^2\dfrac{Z_{01}}{Z_{02}}}$$

$$\tan 2\varphi_2 = \frac{2\left(\dfrac{AB}{Z_{01}} - CDZ_{01}\right)}{C^2 Z_{01}Z_{02} + D^2\dfrac{Z_{01}}{Z_{02}} - A^2\dfrac{Z_{02}}{Z_{01}} - \dfrac{B^2}{Z_{01}Z_{02}}} \tag{3.44}$$

then the overall two-port network at the corresponding reference planes is a *K*-inverter with *K* value given by:

$$K = \sqrt{\frac{Z_{01}}{Z_{02}}} \bullet \sqrt{\frac{1-|S_{11}|}{1+|S_{11}|}} \tag{3.45}$$

The electrical lengths obtained by (3.27) through (3.43) may be negative and therefore not physically realizable as such. However, it is typically possible that such negative lengths are absorbed into the resonators.

3.4 Coupling Matrix Techniques

The concept of the coupling matrix as applied to dual-mode symmetrical waveguide filters was introduced by Atia and Williams in the early 1970s [11]. Since then, the coupling matrix representation of microwave filters has increasingly been adopted for the design; this is particularly so because representing the circuit as a matrix enables a range of associated algebraic operations while maintaining a unique correspondence between the matrix and physical elements. This simplifies the synthesis and also enables it to account for the specific features of each filter element.

In the coupling matrix approach, the prototype electrical circuit consists of a cascade of lumped-element series resonators coupled by transformers. In Figure 3.13, a general form of a two-port bandpass prototype network composed of synchronously-tuned inductively coupled resonators is shown (reproduced from [11]). *L*, *C*, and *R* denote the inductance, capacitance, and resistance, respectively; *i* represents the loop current, e_1 the voltage source, and $C_i = 1$ *F* for $i = 1$ to n, $L_1 = L_n = 1$ *H*, and $L_i = 1/2$ *H* for $i = 2$ to $n-1$. The resonators are assumed to be lossless, with a normalized resonant frequency of $\omega_0 = 1$ rad/sec. The interconnection (or coupling) represents a means by which the electromagnetic energy can be exchanged between the different resonators. It is expressed by M_{ij}, which is the coupling coefficient between the *i*th and *j*th resonators. In this prototype circuit, only magnetic coupling coefficients are considered and are normalized to a bandwidth (BW) of 1 rad/sec. Similar circuits apply for networks with capacitive couplings [4].

Using Kirchhoff's voltage law, the algebraic sum of the voltage drops around any closed path in a network is zero. In view of Figure 3.13, the loop equations take the form:

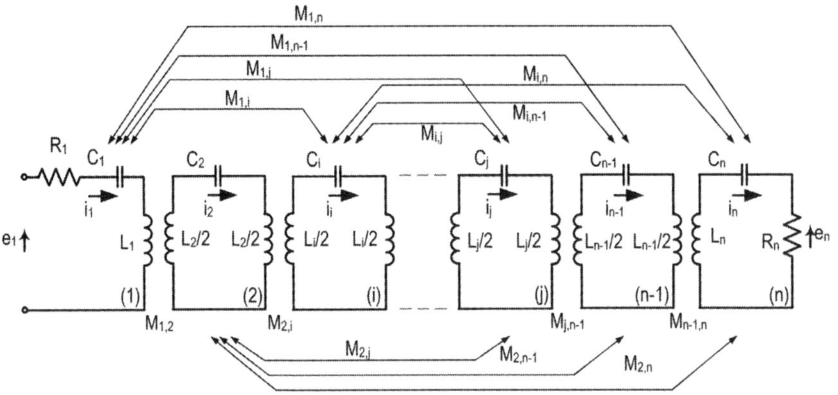

Figure 3.13 A general two-port network composed of synchronously tuned coupled resonators. (Copyright 2014 IEEE [12]).

$$
\begin{aligned}
(R_1 + j\omega L_1 + 1/j\omega C_1) \bullet i_1 + j\omega L_{12} \bullet i_2 + \cdots + j\omega L_{1n} \bullet i_n &= e_1 \\
j\omega L_{21} \bullet i_1 + (j\omega L_2 + 1/j\omega C_2) \bullet i_2 + + \cdots + j\omega L_{2n} \bullet i_n &= 0 \\
&\vdots \\
j\omega L_{n1} \bullet i_1 + j\omega L_{n2} \bullet i_2 + + \cdots + (R_n + j\omega L_n + 1/j\omega C_n) \bullet i_n &= 0
\end{aligned}
\tag{3.46}
$$

in which $L_{kl} = L_{lk}$ represents the mutual inductance between resonators k and l, R_1, R_n are the source and load resistances, is the induced voltage in the loop k due to the current in loop l. Defining the coupling coefficients as $M_{kl} = L_{kl}/L$, it can be assumed that M_{kl} are real (positive, negative or zero) numbers whose values are independent of frequency. The frequency-invariant assumption is valid for narrow bandwidths. In this section, we will consider the case where power exchange between resonators is based on mutual inductance L_{kl} and hence the associated couplings are of magnetic nature. The treatment of capacitive couplings can be found in [4].

The set of equations (3.46) can be represented in matrix form

$$[Z] \times [i] = [e] \tag{3.47}$$

where $[Z]$ is:

$$\begin{bmatrix} R_1 + j\omega L_1 + 1/j\omega C_1 & j\omega L_{12} & \cdots & j\omega L_{1n} \\ j\omega L_{21} & j\omega L_2 + 1/j\omega C_2 & \cdots & j\omega L_{2n} \\ \vdots & \vdots & \ddots & \vdots \\ j\omega L_{n1} & j\omega L_{n2} & \cdots & R_n + j\omega L_n + 1/j\omega C_n \end{bmatrix} \tag{3.48}$$

For a filter with all resonators tuned at the same frequency $\omega_o = 1/\sqrt{LC}$ (synchronously tuned) $[Z]$ can be expressed as:

$$[\bar{Z}] = \omega_0 \times L \times FBW \times [Z] \tag{3.49}$$

where $[\bar{Z}]$ is defined as

$$\begin{bmatrix} \frac{R_1}{FBW\omega_o L} + \frac{j}{FBW}\left(\frac{\omega}{\omega_o} - \frac{\omega_o}{\omega}\right) & j\frac{\omega}{\omega_o}\frac{1}{FBW}\frac{L_{12}}{L} & \cdots & j\frac{\omega}{\omega_o}\frac{1}{FBW}\frac{L_{1n}}{L} \\ j\frac{\omega}{\omega_o}\frac{1}{FBW}\frac{L_{21}}{L} & \frac{j}{FBW}\left(\frac{\omega}{\omega_o} - \frac{\omega_o}{\omega}\right) & \cdots & j\frac{\omega}{\omega_o}\frac{1}{FBW}\frac{L_{2n}}{L} \\ \vdots & \vdots & \ddots & \vdots \\ j\frac{\omega}{\omega_o}\frac{1}{FBW}\frac{L_{n1}}{L} & j\frac{\omega}{\omega_o}\frac{1}{FBW}\frac{L_{n2}}{L} & \cdots & \frac{R_n}{FBW\omega_o L} + \frac{j}{FBW}\left(\frac{\omega}{\omega_o} - \frac{\omega_o}{\omega}\right) \end{bmatrix} \tag{3.50}$$

For a narrowband filter we can approximate $\omega/\omega_0 \approx 1$ and therefore write

$$j\frac{\omega}{\omega_o}\frac{1}{FBW}\frac{L_{kl}}{L} \approx \frac{jM_{kl}}{FBW} = jm_{kl} \tag{3.51}$$

where m_{kl} is the normalized coupling coefficient. It should be noted that the external quality factor, Q_e, is defined as

$$Q_{ek} = \frac{\omega_o \bullet L}{R_k} \ for \ k = 1, n \tag{3.52}$$

and the normalized external quality factors, qe, as

$$q_{ek} = Q_{ek} \bullet FBW \ for \ k = 1, n \tag{3.53}$$

Hence (3.47) takes the form:

$$(\lambda I - jR + M) \bullet i = -je \tag{3.54}$$

where R is an $n \times n$ matrix whose elements are all zero (except for the (1, 1) and (n, n) entries, which are R_o and R_n, respectively), I is the $n \times n$ identity matrix, M is the nxn coupling matrix, i is the $n \times 1$ current vector, e is the $nx1$ excitation vector with all its elements zero except for $e1$, and λ defined by:

$$\lambda = \frac{1}{FBW} \bullet \left(\frac{\omega}{\omega_o} - \frac{\omega_o}{\omega} \right) \tag{3.55}$$

3.4.1 Analysis

The analysis problem involves the calculation of the currents when the elements of the coupling matrix M are known. In order to do this, the system of equations is solved for the currents, and the transfer voltage ratio ($i_n R_n / e_1$) can be determined. This is achieved by using the inverse of the matrix ($\lambda I - jR + M$) or [Z]. The currents can then be found:

$$i = -j(\lambda I - jR + M)^{-1} \bullet e \quad or \quad i = -j[Z]^{-1} \bullet e \tag{3.56}$$

where [Z] is a matrix of impedances. By inspecting the n-coupled resonator network of Figure 3.13 and the two-port reactance network (Figure 3.14) it can be identified that $I_1 = i_1$, $I_2 = -i_n$, $V_1 = e_1 - I_1 R_1$, $V_2 = i_n R_n$. Considering the standard definitions for the short-circuit admittance matrix [y],

$$\begin{bmatrix} I_1 \\ I_2 \end{bmatrix} = \begin{bmatrix} y_{11} & y_{12} \\ y_{21} & y_{22} \end{bmatrix} \bullet \begin{bmatrix} V_1 \\ V_2 \end{bmatrix} \tag{3.57}$$

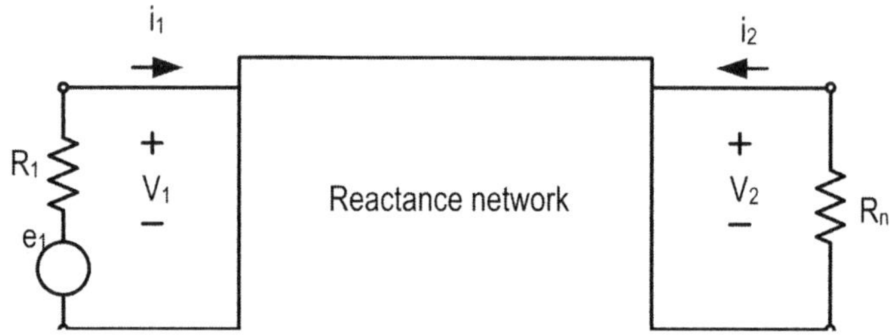

Figure 3.14 A general two-port network composed of synchronously tuned coupled resonators. (Reproduced from [11].)

the two-port y parameters can be obtained in terms of the coupling matrix, M, and the frequency variable, s, by solving (3.54) with $R = 0$ and $e_1 = 1$:

$$\begin{aligned} y_{11}(\lambda) &= -j\left[(\lambda I + M)^{-1}\right]_{11} \\ y_{21}(\lambda) &= -j\left[(\lambda I + M)^{-1}\right]_{nn} \end{aligned} \tag{3.58}$$

Given the coupling matrix, $[M]$, the filter S-parameters are obtained by:

$$\begin{aligned} S_{11} &= 1 + 2jR_1\left[(\lambda I - jR + M)^{-1}\right]_{11} \\ S_{21} &= -2j\sqrt{R_1 R_n}(\lambda I - jR + M)^{-1} \end{aligned} \tag{3.59}$$

3.4.2 Synthesis

The synthesis problem consists of formulating the two-port, short-circuit admittance parameters, y, in two ways: from the coefficients of the known characteristic polynomials that define the desired transfer and reflection characteristics S_{21} and S_{11}; and from the elements of the coupling matrix itself. By equating the two formulations, the values of the coupling matrix are related to the coefficients of the transfer and reflection polynomials [4, 11]. The formulation of the coupling matrix from the lowpass prototype circuit elements is presented in [6, Section 8.1.3]. For Chebyshev filters, the coupling elements $M_{n;n+1}$ are related to the lumped element prototype elements g_k as follows [4]:

$$M_{j,j+1} = \frac{1}{\sqrt{g_j g_{j+1}}}, \quad \mathit{for}\ j = 1,2,\ldots,N-1$$
$$R_1 = \frac{1}{g_0 g_1},\ R_N = \frac{1}{g_N g_{N+1}} \tag{3.60}$$

Equation (3.60) also illustrates the equivalence of the circuit model with the coupling matrix approach. It is noted that this equivalence is not limited to Chebyshev filtering functions, but is valid for generalized Chebyshev filtering functions as well.

The design of transfer functions and the coupling matrix synthesis of filter networks with more advanced performance characteristics is a matter of recent and ongoing research. The case of filters with asymmetric characteristics is discussed in [13], while the coupling matrix synthesis with multiple input/output couplings is reported in [14]. Techniques for transformation and reduction of the coupling matrix that simplify the physical realization of a filter are discussed in [4]. The treatment of the aforementioned topics extends beyond the scope of this book, and the reader is referred to the relevant literature.

References

[1] Matthaei, G. L., L. Young and E. M. T. Jones, *Microwave Filters, Impedance-Matching Networks and Coupling Structures*, Norwood, MA: Artech House, 1980.

[2] Rhodes, J., *Theory of Electrical Filters*, London: John Wiley and Sons, 1976.

[3] Uher, J., J. Bornemann, and U. Rosenberg, *Waveguide Components for Antenna Feeding Systems: Theory and CAD*, Norwood, MA: Artech House, 1993.

[4] Cameron, R., C. Kudsia, and R. Mansour, *Microwave Filters for Communication Systems: Fundamentals, Design and Applications*, Hoboken, NJ: Wiley-Interscience, 2007.

[5] Hong, J., and M. Lancaster, *Microstrip Filters for RF/Microwave Applications*, Hoboken, NJ: Wiley-Interscience, 2001.

[6] Hunter, I., *Theory and Design of Microwave Filters*, London: the Institution of Electrical Engineers, 2001.

[7] Darlington, S., "A history of network synthesis and filter theory for circuits composed of resistors inductors and capacitors," *IEEE Transactions on Circuits and Systems I: Fundamental Theory and Applications,* Vol. 46, No. 1, pp. 4–13, 1999.

[8] Pozar, D., *Microwave Engineering,* Hoboken, NJ: John Wilely & Sons, 2005.

[9] Cohn, S., "Direct-Coupled Resonator Filters," *Proceedings of the IRE,* Vol. 2, pp. 187–196, 1957.

[10] Levy, R., "A generalized design technique for practical distributed reciprocal ladder networks," *IEEE Transactions on Microwave Theory and Techniques,* Vol. 21, pp. 519–526, 1973.

[11] Atia, A. E., A. E. Williams, and R. Newcomb, "Narrow-band multiple-coupled cavity synthesis," *IEEE Transactions on Circuits and Systems,* Vol. 21, No. 5, pp. 649–655, 1974.

[12] Williams, A. E., and E. Atia, "Narrow-bandpass waveguide filters," *IEEE Transactions on Microwave Theory and Techniques,* Vol. 20, pp. 258–265, April 1972.

[13] Cameron, R., "General coupling matrix synthesis methods for Chebyshev filtering functions," *IEEE Transactions on Microwave Theory and Techniques,* Vol. 47, No. 4, pp. 433–442, 1999.

[14] Cameron, R. J., "Advanced coupling matrix synthesis techniques for microwave filters," *IEEE Transactions on Microwave Theory and Techniques,* Vol. 51, No. 1, pp. 1–10, 2003.

4

Analysis and Design of Helical Resonators

4.1 Introduction

An RF/microwave resonator is a structure that can store electromagnetic (EM) energy at one or more characteristic frequencies. The resonant frequency is the frequency at which the energy stored in the electric field; it equals the energy stored in the magnetic field, and hence, an electromagnetic oscillation can be sustained even in the absence of a continuous external excitation. Resonance is therefore an oscillation phenomenon; one common feature to all resonators is that resonance (i.e., oscillation of electromagnetic energy) can occur in particular oscillation frequencies, referred to as resonance frequencies. Due to the characteristic frequency associated with their resonance, electromagnetic resonators inherently perform as frequency selective elements. Because the phenomenon of resonance occurs in specific frequencies for particular physical configurations, frequency components of a signal in the vicinity of the resonant frequency can trigger resonant oscillations in the resonator. Thus, in the case of a two-port bandpass resonator, the signal incident on one port of the resonator can be picked up at the other port. On the other hand, frequency components away from resonance peter out rapidly within the resonator; these cannot be picked up at the other port but are instead reflected. This mechanism forms the basis on which electrical filters operate; signals in the passband propagate through the filter while signals in the stopband are reflected. Thus, electromagnetic resonators are essential filter elements in many configurations.

The design and performance evaluation of electromagnetic resonators is critical in filter development. Principal design parameters pertaining to the electrical performance of microwave resonators are the resonant frequency, the quality factor Q (Q-factor), and the spurious performance, as well as the power-handing capability. The Q-factor describes the rate of energy loss relative to the stored energy in the resonator while the spurious performance describes higher frequencies where higher order resonances are supported.[1] Early helical resonator design guidelines are presented in [1]. Empirical equations for the dimensions of the helix and cavity are given in [1–3], which are accurate when several constraints are applied.[2] A thorough investigation of filters with helical resonators is presented in [3], where the electromagnetic field distribution in the resonators is provided along with the coupling mechanisms and several issues relating to helical resonator filter technology.

This chapter aims to provide an introduction to theory and techniques suitable for the analysis and design of helical resonators using modern computer-aided design (CAD) tools, such as Ansoft, high frequency structure simulator (HFSS) [4], or computer simulation technology (CST) microwave studio [5]. The helical resonator is initially considered as a particular type of coaxial resonator that supports slow waves by virtue of its helical core. This approach provides valuable physical insight relating to the operation of helical resonators and enables a synthesis procedure that offers increased design flexibility. Several practical design examples that illustrate the application of modern electromagnetic based simulators are subsequently presented. Performance characteristics and design guidelines for helical resonators to assist in the synthesis of helical resonators are provided. The electrical performance of helical resonators is investigated, and a detailed study of the Q-factor that provides design guidelines is presented. The use of tuning elements, as well as dielectric materials in helical resonators, is discussed together with their effect on the performance. Power handling aspects of helical resonators are thoroughly investigated in Chapter 7. Mechanical and thermal properties are discussed in this and in following chapters. A review of the available techniques

1. Higher-order resonances are often unwanted in filters because they can compromise the far out-of-band stopband performance.
2. For example, for circular cross section cavities, the length of the shield in [2] is extended beyond the coil on each side by approximately one quarter of the shield diameter. In the case of a square cross section, the length of the cavity is extended by 30% of the shield length. If this is not the case, the lower few turns of the helix are ineffective for storage of energy but will still contribute to loss [2].

to calculate and measure the performance of microwave resonators in various technologies (e.g., planar, waveguide, coaxial) is given in [6].

4.2 Helical Resonator Considerations

Helical resonators consist of a metallic solenoid enclosed in a shield made from a highly conductive material. Typically, one end of the coil is connected electrically and mechanically at the side or bottom wall of the shield while the other is open-circuited (quarter-wavelength case). The clearance at the top of the resonator reduces the capacitive loading, which would otherwise miniaturize further the resonator. A generic sketch of a helical resonator in a cavity with square cross section is shown in Figure 4.1. Figure 4.2 shows the physical layout of a helical resonator loaded with a tuning element (screw). Other common realizations include cavities with cylindrical cross-section; these will not be considered here because the methods and techniques are directly transferrable. The nomenclature for the dimensions of the helix and cavity—as well as of the coupling irises and tuning screws discussed in the remaining sections of this book—follows the list of symbols provided in Table 4.1 and Table 4.2.

4.2.1 Transmission Lines Model of Resonators

In distributed microwave resonators, it is often convenient to consider energy storage in the form of a superposition of a forward and a backward propagating wave; once appropriate wavelength and phasing conditions are met, the resonance can be thought of as a standing wave, whose properties are dictated by the distributed structure and the associated boundary conditions. Typical boundary conditions include open and short circuit [7], although practical implementations often include reactive (e.g. capacitive) elements that enable miniaturization or tuning of the resonator. If, for simplicity, short or open boundary conditions are assumed, the aforementioned model suggests that the fundamental resonant mode for a finite section of transmission line will be at a frequency where its length is equal to half of the propagating wavelength.

For boundary conditions imposing an open circuit at the two ends of the transmission line resonator, the voltage at the center of the transmission line is zero at resonance [7]. Placing a short circuit at that point will not affect the resonance properties. On this basis, coaxial cavity resonators comprise

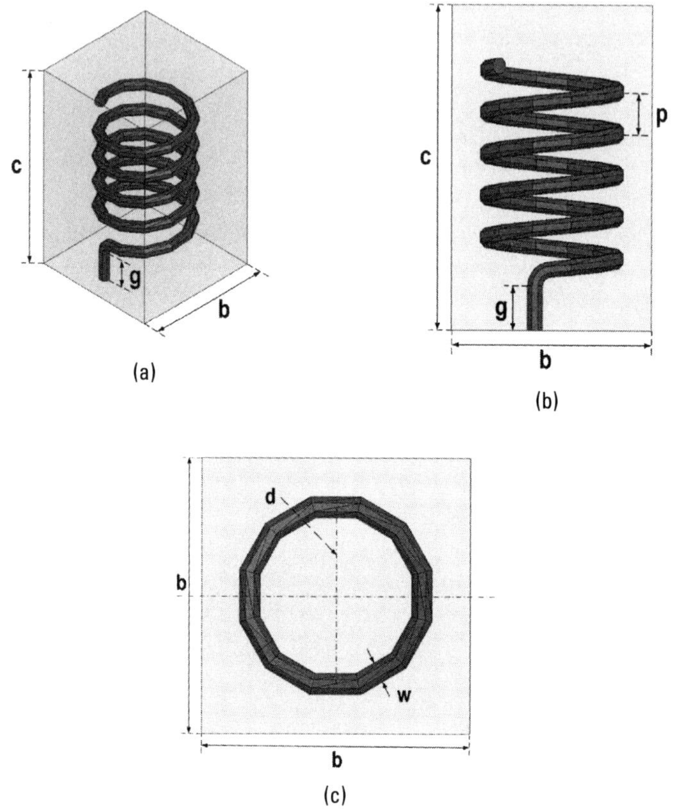

Figure 4.1 Sketch of a single cavity helical resonator and associated symbols for the dimensions (square cross-section cavity enclosure).

a coaxial transmission line of electrical length approximately equal to quarter wavelength, which is open-ended at one end and short-circuited at the other [8]. Similar to half wavelength resonators, knowledge of the propagation characteristics in the transmission line (e.g., propagation constant, wavelength, characteristic impedance) is therefore relevant to the design of this class of resonators.

The electrical length of coaxial resonators can be further reduced by introduction of capacitive loads in the form of additional elements.[3] Likewise, the introduction of tuning screws can add an adjustable capacitive load

3. Although uniform coaxial resonators with dimensions significantly less than one-eighth of the free-space, wavelengths are not commonly encountered in academic literature [8].

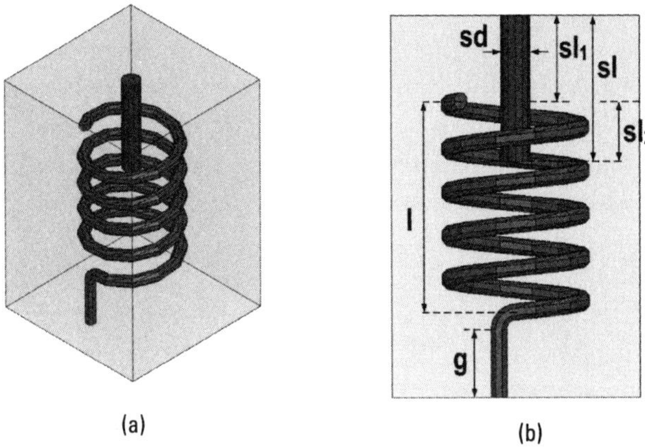

Figure 4.2 Sketch of a single-cavity helical resonator loaded with tuning element (screw). The tuning element penetrates from the top lid of the cavity. It is modeled as a cylinder.

Table 4.1
List of Symbols Used for Helix Cavity.

Helix	Symbol	Cavity	Symbol
Diameter	d	Distance to cavity wall	a_1
Pitch	P	Inter-resonator distance	$a_{12}, a_{23}, \ldots$
Coil axial length	l	Distance to cavity wall	a_2
Wire diameter	w	Length	a
Number of turns	n	Width	b
Ground distance	g	Height	c
Overlap distance	L		

Table 4.2
List of Symbols Used for Tuning Screw and Iris

Tuning Element	Symbol	Iris	Symbol
Diameter	sd	Length	il
Length	sl, sl_1, sl_2	Height	ih
		Width	it

to the resonator that allows for post-fabrication tuning. Even in the absence of intentional reactive loading, the fringing capacitance associated with a physically open-ended distributed structure reduces the electrical length of the resonator. As discussed in Chapter 3, it is also often the case that the coupling mechanism between resonators introduces a negative phich that—when absorbed into the resonator length—can lead to reduction of the electrical length. The effect of these deviations on the resonator performance can be cumbersome to quantify. Yet, the transmission line model for resonators provides both physical insight as well as quantitative information relating to the resonant properties (e.g., [9].)

4.2.2 Coaxial Line with Helical Inner Core

Helical resonators can be seen as a particular type of coaxial cavity resonators short-circuited at one end and open-circuited at the other, such as those being used in coaxial combline and interdigital filters [8, 10]. Unlike traditional coaxial cavity resonators, where the transmission line is a uniform coaxial line, helical resonators comprise a coaxial transmission line with a helical core. The study of the electrical properties of coaxial helical transmission lines can therefore provide useful information about the resonator performance, and, thus, it is beneficial to be performed prior to the design of helical resonators and filters. Coaxial helical transmission lines are periodic structures ore support slow-wave propagation [11]. Significant properties of propagation in periodic structures sed summarized in the dispersion diagram (as in Figures 4.4–4.6), which provides the relationship between the propagation constant, β, and the free-space wave-number, k_0. From this diagram, the design engineer can extract useful information relating to the electrical length and propagation losses at the coaxial helical line.

The dispersion of a coaxial helical transmission line can be rigorously derived by exploiting the eigenmode solver of a full-wave commercial EM software. A unit cell is introduced in the EM eignemode solver, and periodic boundary conditions (sometimes referred to master-slave Floquet boundary conditions) are enforced at the two boundary surfaces that define the period and their separation corresponds to the helix pitch. These boundary conditions impose that the field distributions at the two surfaces are identical with the exception of a phase difference, $\Delta\varphi$, between them. The dispersion analysis can be conducted by imposing a range of phase differences, $\Delta\varphi$, in the direction of propagation (i.e., between the two surfaces assigned to periodic boundary conditions) and calculating the eigen-frequency for each phase angle. The propagation constant, β, is calculated from the imposed phase angle, $\Delta\varphi$, divided by the pitch length, p,

$$\beta = \frac{\Delta\varphi}{p} \tag{4.1}$$

The dispersion diagram can therefore be obtained by plotting the propagation constant as calculated versus frequency for different values of $\Delta\varphi$.

An example configuration for the dispersion analysis of a helical transmission line is shown in Figure 4.3, where all parameters necessary to determining the geometry are provided. In the remaining paragraphs of this section, this example is employed to present some general design guidelines, as well as physical insight on the propagation characteristics. In particular, a coaxial helical transmission line (in the remaining referred to as helix00) with dimensions as in Table 4.3 is considered as reference. The effect of the helix pitch p, the diameter of the helix d, as well as the separation between the cav-

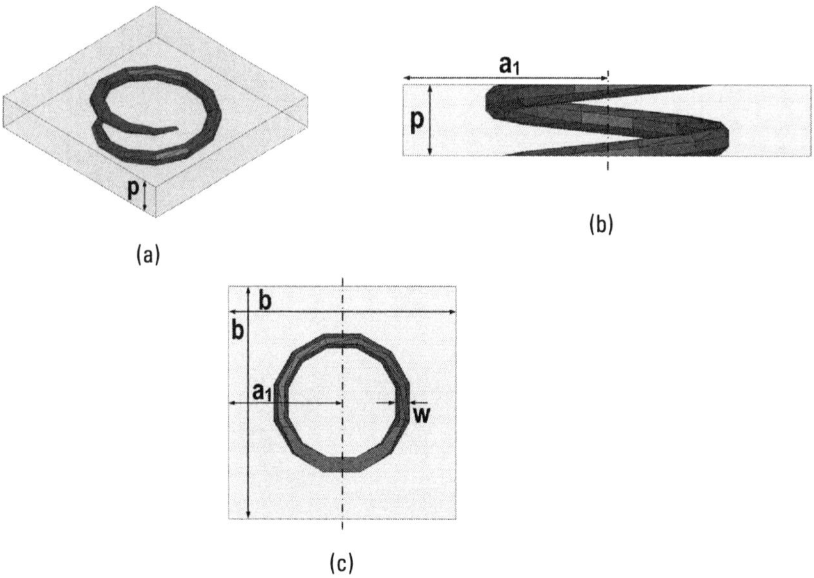

Figure 4.3 Helical transmission line: Configuration of a single turn/pitch used for the dispersion analysis (helix00).

Table 4.3
Helical Transmission Line: Helix and Side Wall Dimensions (Helix00)

Helix	mm	Side Walls	mm
d	21.06	$a_1=a_2$	19.15
p	6.53	*b*	38.30
w	2.62		

ity and the side walls (which, in turn, depends on the width of the cavity *b*) are then investigated by means of a set of parametric studies.

Figure 4.4 shows the dispersion for the reference line for varying pitch length, *p*. Reduced pitch length results in greater value for the wave-number, β, and consequently a shorter propagating wavelength, λ_g. This can be attributed to the associated longer path per period for the current flowing at the core of the coaxial line. Figure 4.5 shows the effect of the cavity side wall width, *b*, on the dispersion relation. According to Figure 4.5, as the cross-sectional dimension of the cavity approaches the diameter of the helix, the

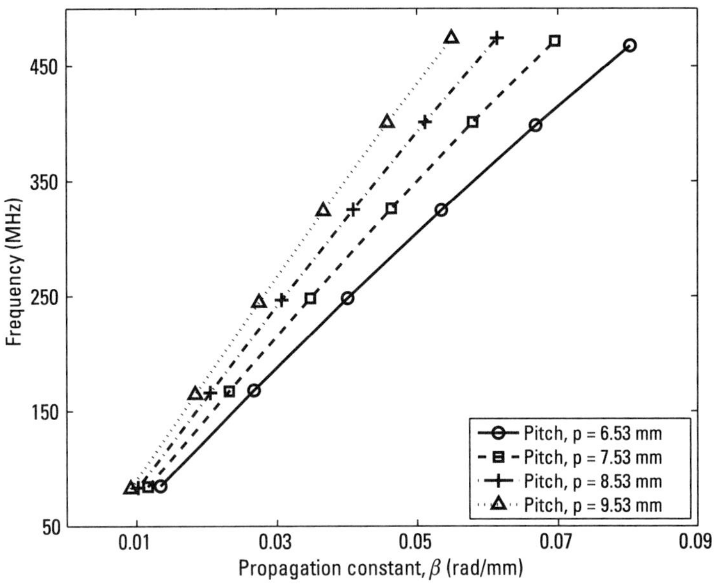

Figure 4.4 Helical transmission line (helix00): Dispersion curves, parametric study with the length of the pitch.

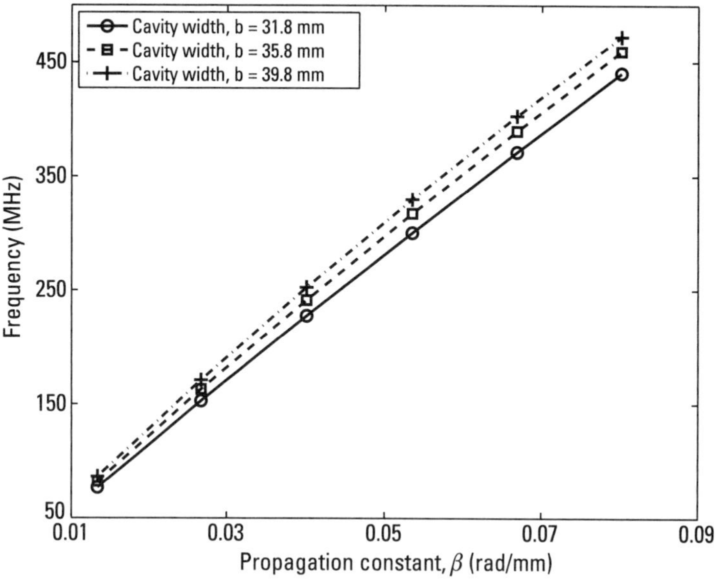

Figure 4.5 Helical transmission line (helix00): Dispersion curves, parametric study with the cavity coss-section.

wave-number, β, is increased and the propagating wavelength, λ_g, is reduced. This miniaturization can be attributed to the higher capacitive loading per unit length of the transmission line. Figure 4.6 shows the dispersion of a helical transmission line with varying helix diameter. The effect of the diameter is by far more pronounced on the wave-number when compared to the effect of the pitch and the cavity cross section; this is attributed to the substantial increase in current path provided for larger helix diameter.

4.2.3 Helical Coaxial Transmission Line Resonators

As discussed above, the reduced physical dimensions of helical resonator filters are achieved by virtue of the slow-wave properties of helical transmission lines and the associated wavelength shrinkage [11]; as a result of the reduction of the propagating wavelength, a fixed electrical length corresponds to reduced physical dimensions. An example of a helical resonator formed by the coaxial lines studied in Section 4.2.2 is shown in Figure 4.1. In this case, the helix is grounded at its lower end to the bottom of the cavity; alternative configurations entail electrical connections between the helix and the cavity at one of the cavity's side walls and have similar characteristics. It is noted that

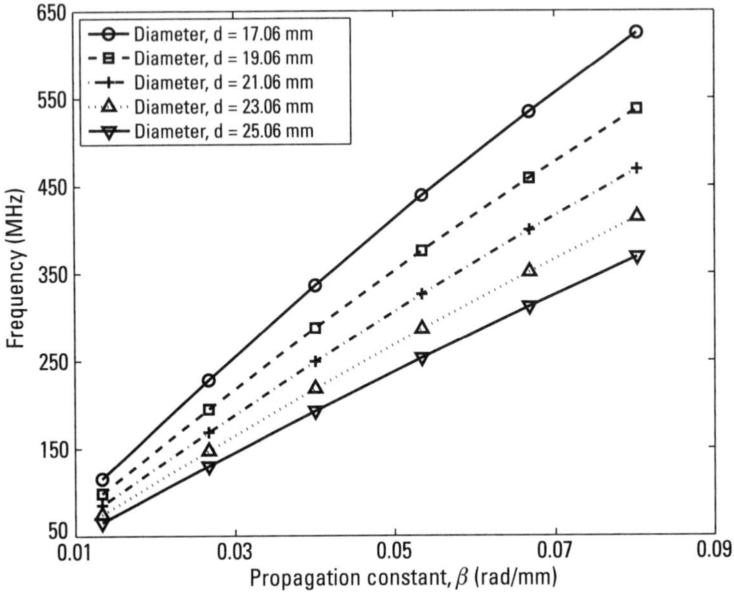

Figure 4.6 Helical transmission line (helix00): Dispersion curves, parametric study with the helix diameter.

the technique presented in this section can be extended to helical resonators with modulated radius and therefore varying wavelength along their axis [12].

The application of the dispersion analysis in the design of helical resonators is illustrated by means of an example involving the design of four resonators based on the reference design (helix00) with layout as shown in Figure 4.1, each with a different pitch, p. It is required that all resonators operate at the same frequency, f_0, here selected to be 300 MHz. Using the method described in Section 4.2.2, the propagation constant is estimated for the different pitch values at the operating frequency. This enables to estimate the guided wavelength, λ_g, in each case. Initial estimation for the resonant physical length can therefore be obtained by considering electrical length in each case equal to $\lambda_g/4$. The physical axial length of each resonator as estimated following this procedure is shown in Table 4.4.

The shield is extended on each side of the transmission line by the same distance in all four cases. The bottom and top of the shield are covered, and the helix is grounded at the bottom end to form the four resonators (see Table 4.4 for dimensions). The resonant frequency for all four resonators as calculated using commercial full wave eigenmode solvers is found to be

Table 4.4
Design Example of Four Resonators Based on Dispersion Analysis*

Pitch (mm)	β@300 MHz(rad/mm)	$\lambda_g/4$ (mm)	N	c (mm)
p_1=6.53	0.04895	32.091	4.916	57.622
p_2=7.53	0.04238	**37.065**	4.924	62.596
p_3=8.53	0.03750	41.888	4.912	67.434
p_4=9.53	0.03387	46.383	4.869	71.914

*The dimensions other than helix pitch, p are as in Table 4.3.

approximately at 272 MHz, and is almost 9.3% lower than the target 300 MHz. The fact that all four resonators operate at the same frequency validates the use of the dispersion curve as a means to design their geometrical parameters. Some numerical optimization is required to account for the fringing effects[4] that are dependent on the resonator geometry and frequency. It is further noted that rotating the helix around its axis has a small effect on the resonant frequency. Calculations of the resonant frequency obtained by an eigenmode solver reveal that rotation of the helix in steps of 45° around its axis leads to maximum fractional variation of the resonant frequency of less than 0.5%.

4.3 Full-Wave Modeling of Helical Resonators

The electromagnetic modeling of helical resonators provides not only an accurate estimation of the resonant frequency, but also other useful information, such as the *Q*-factor and the field distribution. For resonators that are arbitrarily shapd, (and therefore analytical or semianalytical modeling becomes cumbersome), full-wave EM simulators are commonly employed for the accurate calculation of the resonant frequency. Recent progress in full-wave 3-D EM analysis allows for precise investigation of the electrical properties of helical resonators. Either an eigenmode analysis or a driven *S*-parameter analysis of most EM CAD tools can be used for this purpose. Both

4. According to Zverev [2], for a helical resonator with circular cross section, the combination of the fringing effects and self-capacitance leads to an electrical length for the coil approximately 6% less than a quarter wavelength.

techniques can provide the resonant frequency of the fundamental mode as well as the resonant frequencies of the higher-order modes.

Eigenmode solutions directly provide both the unloaded Q-factor as well as the modal field distribution. Driven analysis requires inclusion in the model of an input coupling structure that will excite the resonant structure with the signal at the driven port. Such a coupling mechanism will act as reactive loading to the resonator and will consequently reduce the estimation of the resonant frequency. Attention should therefore be paid to ensure that the external coupling is weak in order to minimize the external loading. The driven solution can provide both the unloaded Q-factor and the external Q-factor. Is should be noted that driven analysis will only provide the resonant frequency of the modes that the specific excitation structure (e.g., conductive probe, inductive loop, or capacitive tap) can excite. Attention should therefore also be paid in ensuring suitable excitation strategies in driven analysis of helical resonators. Chapter 5 provides further details on practical realization of both techniques and practical excitation methods.

In Figure 4.7, a configuration suitable for a driven S-parameters calculation of the resonant frequency is shown. A 50-ohm SMA coaxial line is used as an input port. At its other end, the inner conductor is extended and connects to the helix (conducting probe coupling), thus implementing inductive coupling. The mechanical and electrical connection method is dependent on the materials of the helix conductor and SMA. Weak external coupling conditions can be achieved by normalizing the driven port input impedance to a

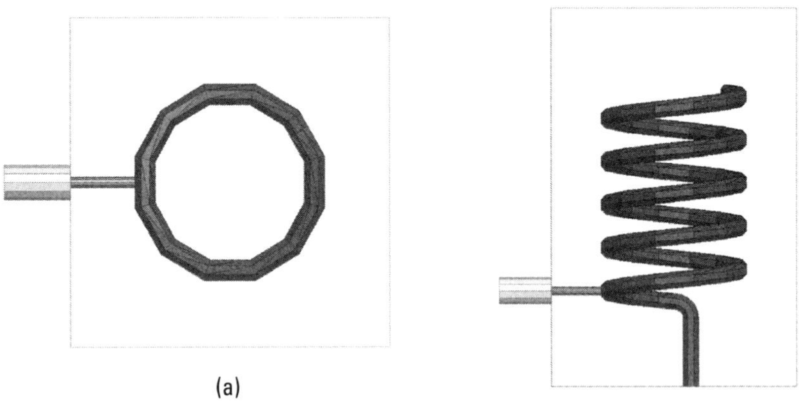

(a) (b)

Figure 4.7 Model of a helical resonator with input tap coupling (SMA coaxial line).

value that ensures a high level of mismatching. Using a driven solver and accounting for Ohmic loss, the input reflection coefficient S_{11} at the waveguide port can then be derived versus frequency. The location of the S_{11} minimum indicates the frequencies where the energy is partially stored in the resonator [6]; at other frequencies, almost all the energy is reflected back.

The application of eigenmode and driven solvers for the analysis of helical resonator filters is demonstrated by means of a resonator that exploits the reference helical coaxial transmission line of Table 4.3.The dimensions of the resonator are provided in Table 4.5. Table 4.6 gives the calculated values of the resonant frequencies for the first three modes through an eigenmode analysis. As shown, the fundamental resonance occurs at 275 MHz. Figure 4.8 shows the reflection coefficient for the same resonator estimated by a driven solution (conductive probe coupling) upon weak external excitation. The resonance frequency is marked at 275 MHz, which is identical to that calculated with eigenmode analysis and thus verifies that the two techniques give similar results. The resonant frequency of the second mode is calculated

Table 4.5
Helical Resonator (Helix00): Helix and Cavity Dimensions*

Helix	mm(λ_0)	Cavity	mm(λ_0)
d	21.06	a_1=a_2	19.15
p	6.53	b	38.3(.0351)
l	31.64(.0274)	c	57.4(.0527)
w	2.62		
n	4.85		
g	10.29		

*Lengths normalized to freespace wavelength are indicated in parentheses.

Table 4.6
Helix00—First Three Resonant Modes*

Resonant Mode	Frequency (MHz)	Quality Factor, Q
1	275	1454
2	767	1940
3	1200	1943

*Dimensions given in Table 4.5

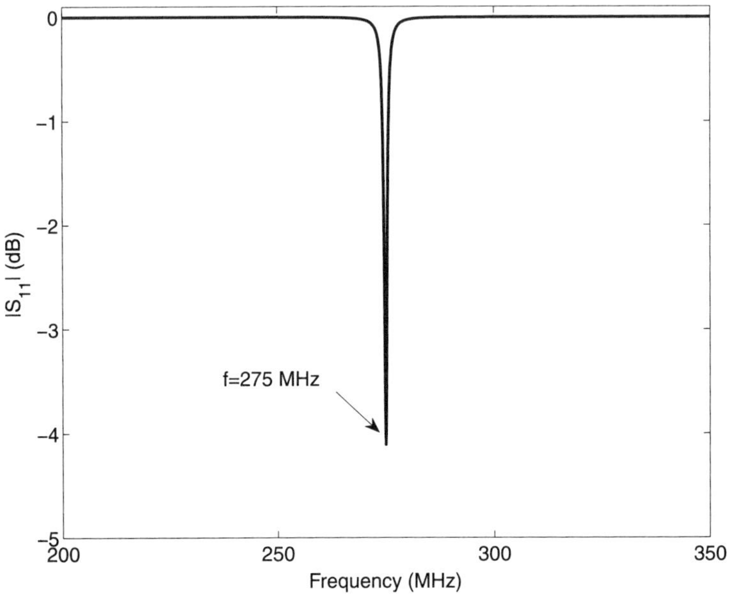

Figure 4.8 *S*-parameters ($|S_{11}|$) response obtained from the weak coupling analysis corresponding to helix00. Simulation accounts for ohmic loss.

at 767 MHz, and the spurious-free window in this case is therefore also 492 MHz. The calculated values for the unloaded Q_0 of these modes are also provided in Table 4.6 (see Section 4.4).

According to the transmission line analogy discussed in Section 4.2.3, a $\lambda_g/4$ resonator short-circuited at one end and open-ended at the other will produce a current maximum at the former and a voltage maximum at the latter. Considering the definitions of currents and voltages for distributed transmission lines, this analogy suggests that a magnetic field maximum occurs at the short-circuited end, while an electric field maximum occurs at the open-ended end. This is demonstrated with the use of a generic eigenmode solver that can conveniently calculate the electric and magnetic field distributions at a given resonant mode. The electric and magnetic field distribution for the example of Table 4.5 are shown in Figures 4.9 and 4.10, respectively. With view of Figures 4.9 and 4.10 one can categorize the mode in the helical resonator as quasi-TEM. Figure 4.11 shows the surface current density (magnitude) at the fundamental resonant mode at the surface of the helix and at the surface of the helix and the surrounding walls of the cavity enclosure.

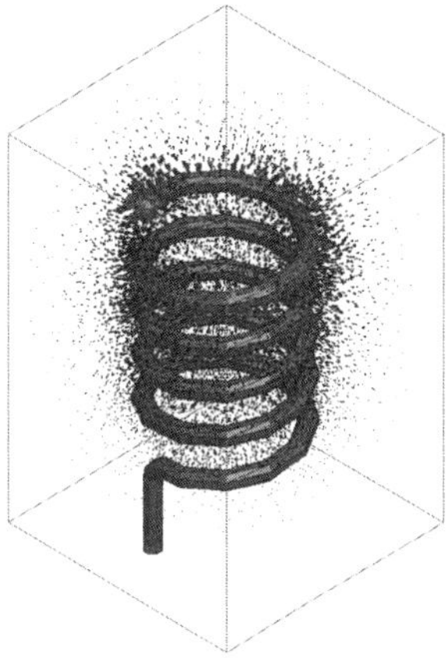

(a)

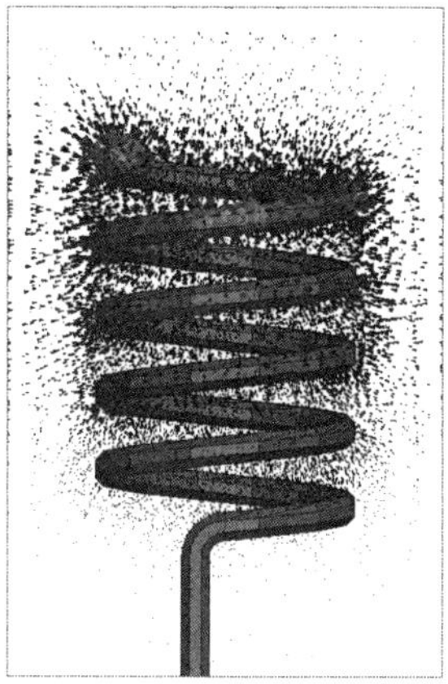

(b)

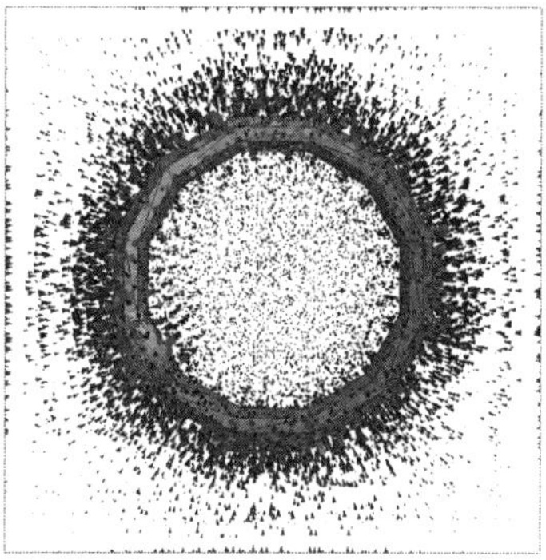

(c)

Figure 4.9 Helical resonator: Field distribution at the fundamental resonant mode—electric (helix 00).

Distributed microwave resonators, unlike idealized lumped LC resonators, can support an infinite number of resonant modes, each occurring at distinct frequency point characterized by a distinct electromagnetic field distribution. The proximity of the neighboring resonant modes to the operating mode defines the spurious performance of microwave resonators. The transmission line analogy predicts that the resonant frequency of the higher-order mode should approximately occur at a value three times that of the fundamental mode. Given that the fundamental mode for helix00

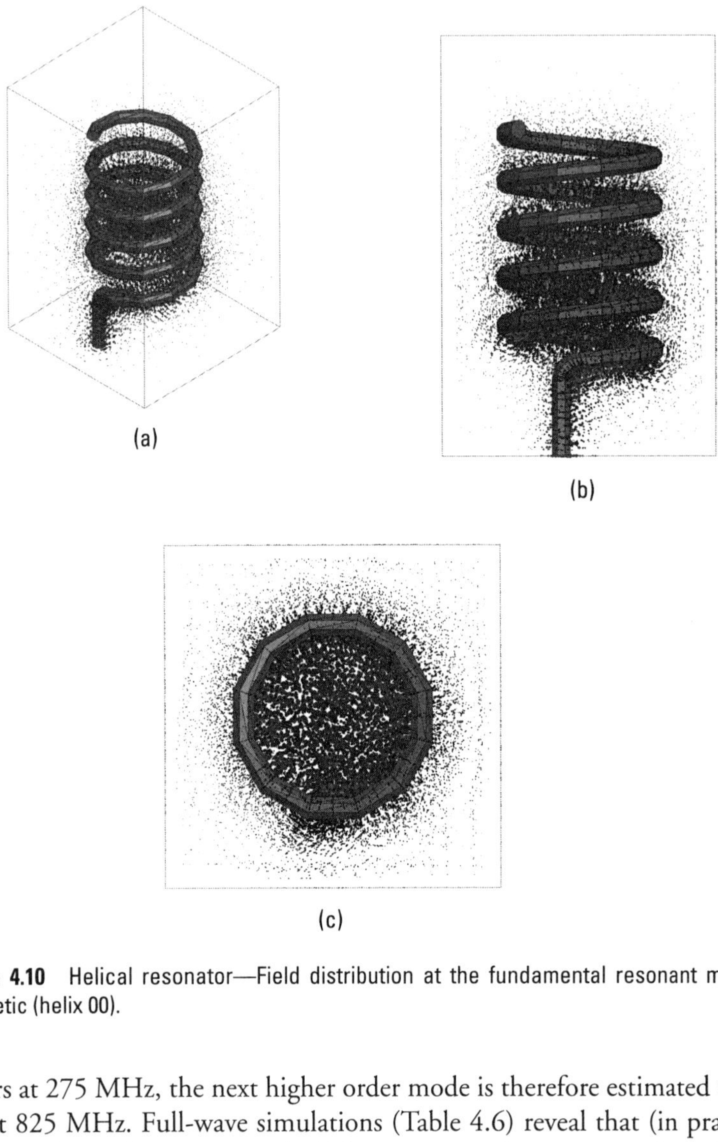

Figure 4.10 Helical resonator—Field distribution at the fundamental resonant mode—magnetic (helix 00).

occurs at 275 MHz, the next higher order mode is therefore estimated to occur at 825 MHz. Full-wave simulations (Table 4.6) reveal that (in practice) higher-order mode occurs at 767 MHz, corresponding to approximately 7% reduction from the former estimation. This is attributed mainly to the resonator cavity walls that present a different level of capacitive loading at the two frequencies (275 and 767 MHz), thus affecting the two resonant frequencies differently. In the case of helical resonators, (and in contrast to uniform coaxial resonators), the discrepancy can also be attributed to the nonlinear

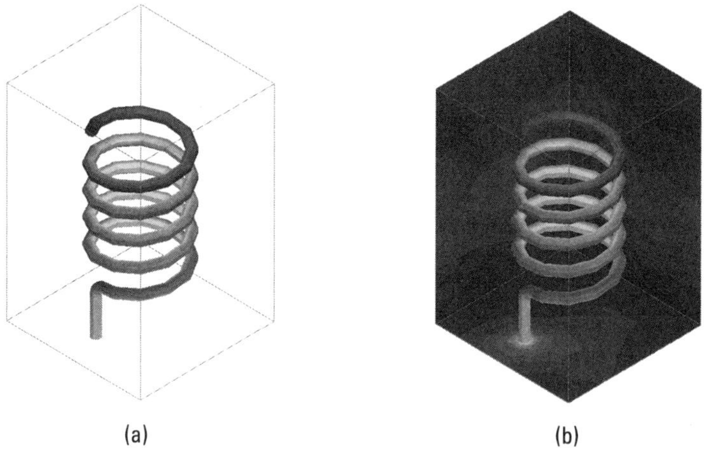

Figure 4.11 Helical resonator: Surface current density (magnitude) at the fundamental resonant mode; (a) at the surface of the helix; (b) at the surface of the helix and the bottom and side walls of the cavity enclosure—(helix 00). Greater intensity represents greater current density.

frequency dispersion of the helical transmission line. It is further noted that the higher-order propagating modes in the coaxial helical transmission line can contribute to additional higher-order modes that are not accounted for by the first-order transmission line model.

4.4 Quality Factor

An important parameter describing the performance of a resonant circuit is its quality factor, Q. It is defined as

$$Q = \omega \frac{\text{average energy stored}}{\text{energy loss per second}} \tag{4.2}$$

Three definitions of the quality factor for a resonator are available, namely the unloaded quality factor, Q_0, the loaded quality factor, Q_l, and the external quality factor, Q_e. The reader is referred to [7] for the formal definitions. Here, we note that the unloaded quality factor characterizes the resonator in isolation and therefore the energy loss in (4.2) is only attributed to thermal losses in the resonator. In contrast, the external quality factor characterizes the coupling of a resonator to its neighboring circuitry, and in this case, the energy loss in (4.3) only considers the energy exchange between the resonator and the remaining circuitry. The loaded quality factor considers both thermal losses within the resonator and the energy exchange between the resonator and its environment. Therefore:

$$\frac{1}{Q_l} = \frac{1}{Q_e} + \frac{1}{Q_0} \tag{4.3}$$

The unloaded quality factor of a resonator, Q_0, is critical in determining the insertion loss of a filter. A key advantage of helical resonators at VHF frequencies and up to the L-band is that they provide a tuned circuit whose unloaded quality factor Q_0 is much higher than that of the practical lumped-element circuits. Possible causes of dissipation in helical resonators are ohmic losses at the helix and the conductive enclosure, as well as any dielectric losses in case additional insulating material is included in the structure (see Section 4.5). Approximate formulas for Q-factor calculation of helical resonators are given in [2]. Rigorous calculation of the unloaded quality factor Q_0 can be readily obtained using eigenmode analysis; each eigenvalue corresponds to a resonant frequency while the corresponding eigenvector corresponds to the field distribution characteristic for this resonant mode. Exploiting the field distribution once the conductivity of all metalization and loss tangent of all dielectric materials have been considered, the Q_0 value can be readily estimated. In practice, most commercial electromagnetic solvers with an eigenvalue function would directly return the unloaded Q_0 value. External and loaded quality factors can be estimated only by a driven solution and from those the unloaded quality factor can be extracted using (4.3).

A set of parametric studies are presented in to provide general design guidelines regarding geometrical parameters that optimize the Q_0 of helical resonators. The conductivity of silver (Au, $\sigma = 6.1 \times 10^{07}$ s/m) has been considered both for the helix and the cavity walls, whereas no dielectric material has been considered. Initially, the effect of the cavity width, b, on Q_0 is presented by virtue of studying the unloaded quality factor for a series of resonators with fixed helix dimensions (helix00) and varying width of the cavity, b. The location of the top and the bottom cavity walls (and therefore the cavity height) remains unchanged. It is noted that—as a result of varying the cavity width—the resonant frequency, f_0, also varies. The calculated f_0 and Q_0 for different values of the cavity width are plotted in Figure 4.12(a).

The resonant frequency f_0 is increased for larger cavity cross section, from a value of 235.2 MHz for $b = 26.3$ mm to almost a value of 287.2 MHz for $b = 48.3$ mm. This 22.1% increase in the resonant frequency is attributed to the reduced capacitance per unit length of the coaxial helical line and is in line with the results presented in Figure 4.5. The variation of the unloaded Q_0 as a function of b suggests that Q_0 is higher for broader cavities. For this example, Figure 4.12(a) reveals a *Q*-factor increase from a value of 816 to a value of 1549 corresponding to 89.8% fractional increase. Figure 4.12(b) plots Q_0 as a function of the cavity width normalized to the free-space wave-length λ_0 at the resonant frequency. The observed trend suggests an almost linear increase of the unloaded quality factor with the normalized cavity width.

A similar study is conducted for the diameter of the helical resonator *d*. The width of the cavity, the number of the helix turns—as well as the pitch (and therefore the physical length) of the helix—remain fixed, while the diameter of the coil is now varied. Figure 4.13(a) plots the resonant frequency, f_0, and unloaded quality factor, Q_0, as a function of the helix diameter, d. In agreement with the conclusions derived from Figure 4.6, the resonant frequency, f_0, drifts significantly towards lower values for larger helix diameters. Figure 4.13(a) reveals a similar trend for the unloaded quality factor; given that the total physical volume of the resonator remains constant, the reduction of the Q-factor for lower resonant frequency is consistent with the general trend of increased thermal losses in physically smaller resonators [13]. The unloaded quality factor, Q_0, versus the normalized length of the resonator is shown in Figure 4.13(b). The trend shows an approximately flat value for Q_0 down to cavities with dimensions of about 0.05 wavelength, below, which the unloaded quality factor reduces rapidly.

It is interesting to note in Figure 4.12 that as the cavity widths, *b*, decreases to values that are close to the circumference of the helix, the resonator *Q*-factor drops significantly. A similar observation can be made in the case of the results obtained in Figure 4.13(a), where the unloaded Q_0 for the helix diameter values *d* that are close to the cavity dimensions *b* again drastically reduces. In order to corroborate these observations, Figure 4.14 shows the unloaded Q_0 of the resonator normalized to frequency and cavity width as a function of the helix diameter *d* over cavity width ratio (*d*/*b*). The results are obtained from the data of the two previous studies. The importance of this result lies in the fact that the unloaded quality factor Q_0 of helical resonators decreases rapidly as the ratio of helix diameter to the width of the cavity (*d*/*b*) deviates away from an optimum value, here approximately 0.–0.65.

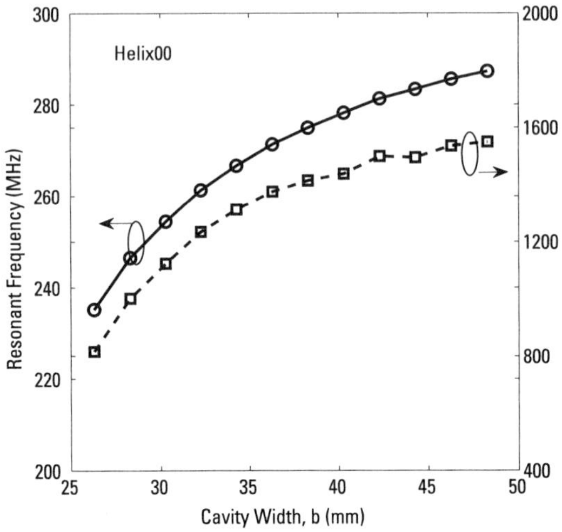

(a)

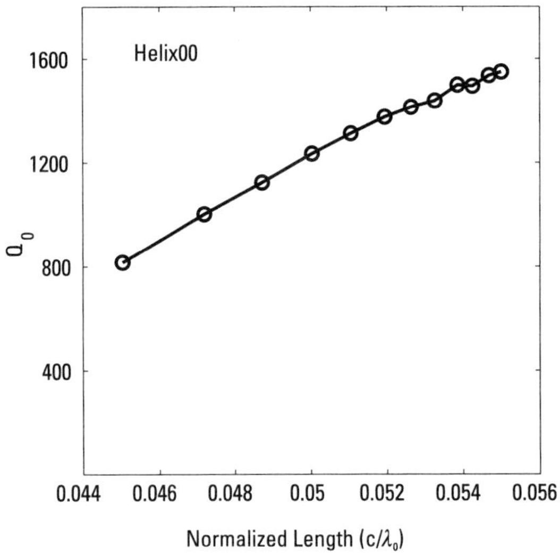

(b)

Figure 4.12 Q_0 of helical resonator. Parametric study of cavity width *b*: (a) Calculated resonant frequency f_0 and Q_0 as a function of the cavity width *b*; (b) Q_0 as a function of the resonators normalized length.

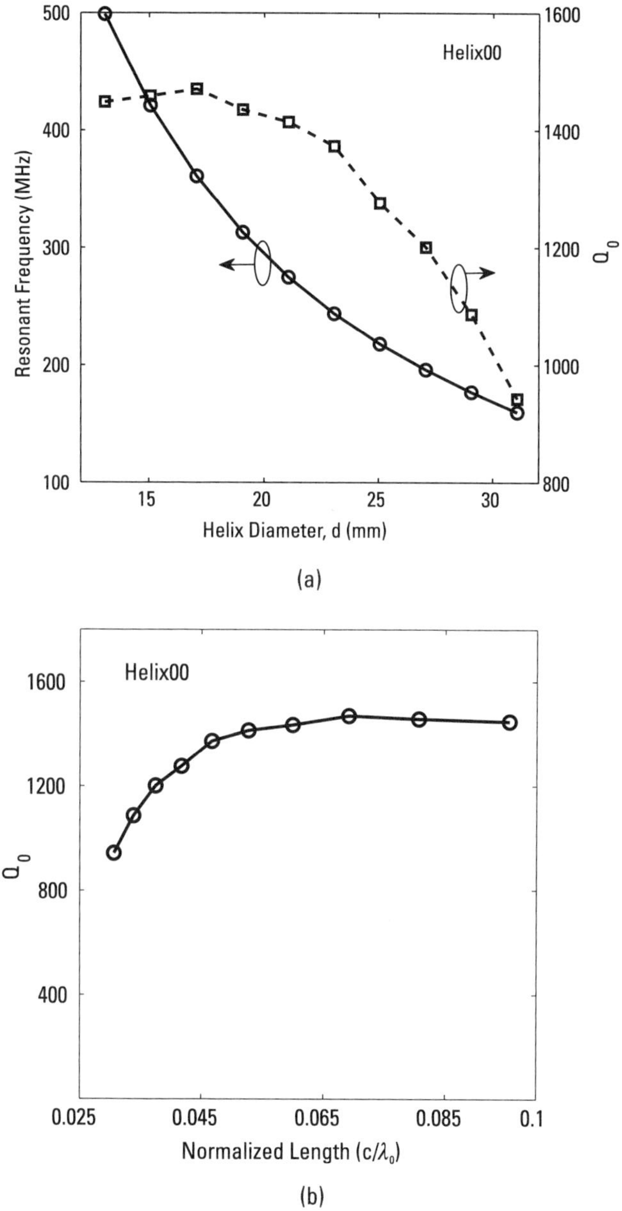

Figure 4.13 Q_0 of helical resonator. Parametric study of helix diameter *d*. (a) Calculated resonant frequency f_0 and Q_0 as a function of the diameter of the helix, *d*. (b) Q0 as a function of the resonator's normalized length.

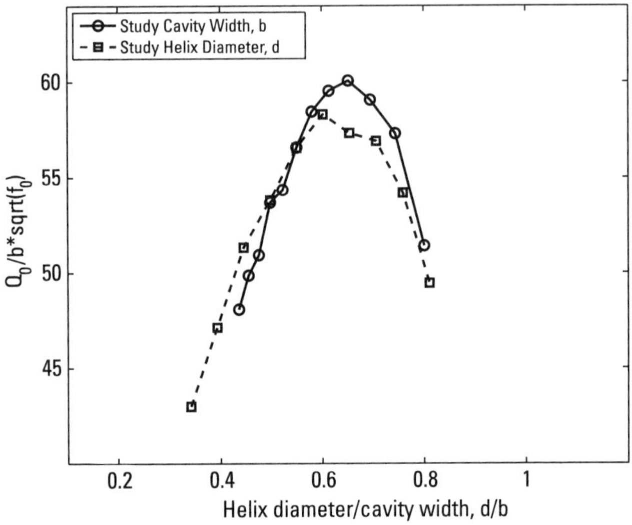

Figure 4.14 Normalized Q_0 of helical resonator as a function of the helix diameter over the cavity width ratio (*d*/*b*). Results obtained from the two previously considered studies.

4.5 Dielectric Loading and Tuning

Dielectric materials and metallic tuning elements are often employed in helical resonators for mechanical support and post-fabrication frequency adjustment, respectively. In this section, we review the effect of adding dielectric materials and metallic tuning elements on the resonator electrical performance.

4.5.1 Dielectric Loading

Qualification protocols in space applications require robustness against mechanical/acoustic vibrations, which in turn pose the need for mechanical support of the helices. The use of dielectric support for the helix has been proposed as a technique to provide additional mechanical support for on-board filters [14]. Likewise, dielectric filling can be used to reduce the resonant frequency and hence the physical size [15, 16]. Figure 4.15 shows a model of a helical resonator example loaded with a hollow dielectric cylinder to provide stability. In order to investigate the effects of the dielectric loading, we employ the helical resonator with dimensions as those of helix00 presented in Section 4.3. Two different dielectric materials are considered, namely the space quali-

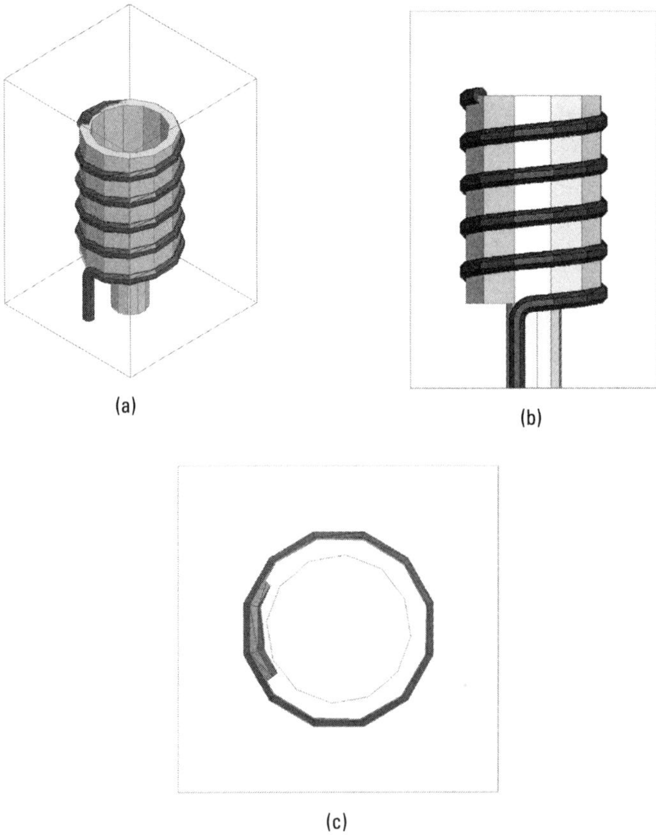

Figure 4.15 Helical resonator loaded with dielectric support frame—example configuration.

fied Rexolite 1422 ($\varepsilon_r = 2.53$ and tanδ 3×10^{-4}) as well as Alumina (Al_2O_3, $\varepsilon_r = 9.8$, tanδ $=4 \times 10^{-5}$). The thickness of the dielectric layer is in both cases 2.6 mm. The calculated values for the resonant frequency and quality factor Q_0 with the inclusion of a 2.6-mm layer of Rexolite are given in Table 4.7. The resonant frequency, f_0, of the first mode drops by about 3.3%, which is attributed to miniaturization by virtue of the dielectric. A 9.5% decrease in the unloaded quality factor, Q_0, is likewise attributed to the dielectric losses. The field distribution inside the cavity is very similar to the one without the dielectric and thus is not shown. Table 4.8 reports the obtained results for alumina and similar trends are observed.

Table 4.7
First Three Resonant Modes: Dielectric-Loaded Helix (Helix 00/ Rexolite 1422)—Dielectric Thickness 2.6 mm

Resonant Mode	Frequency (MHz)	Quality Factor, *Q*
1	266 (275)	1315 (1454)
2	692	1498
3	1038	1390

Table 4.8
First Three Resonant Modes: Dielectric-Loaded Helix (Helix 00/AlO3 Ceramic—Alumina)—Dielectric Thickness 2.6 mm

Resonant Mode	Frequency (MHz)	Quality Factor, *Q*
1	241 (275)	1316 (1454)
2	511	1486
3	711	1367

4.5.2 Tuning

Top loading of the resonator with metallic screws/caps is a commonly employed technique for the post-fabrication tuning of coaxial and helical resonator filters. An example of a helical resonator with a tuning screw is shown in Figure 4.2. The screw is inserted from the top wall of the cavity towards the open end of the helix in the direction of the helix axis. The position of the screw relative to the helix determines an additional capacitance loading, and can therefore be exploited to set the resonant frequency for each resonator. It is noted that tuning screws are not only limited to the control of the resonant frequency but can also be exploited to tune the coupling between successive resonators. For example, in the case of helical resonators coupled through an iris, the height of a tuning screw penetrating the cavity wall at the plane of the iris can be adjusted to control the coupling. The effect of such a tuning screw has not been considered here, although similar techniques can be employed.

The effect of introducing the tuning element of Figure 4.2 on both the resonant frequency and the unloaded quality factor, Q_0, is reported in Table 4.9 for the case of a purely metallic element and in Table 4.10 for the case of a dielectric element (Rexolite 1422). The dimensions of the screws are provided in the legends. In the former case, the resonant frequency is reduced

Table 4.9
Firse Three Resonant Modes: Tuning Screw (Helix 00)—Screw: Silver (Au), *sd* = 6.4 mm, *sl* = 20 m

Resonant Mode	Frequency (MHz)	Quality Factor, *Q*
1	252 (275)	1245 (1454)
2	766	1764
3	1201	1750

Table 4.10
First Three Resonant Modes: Tuning Screw (Helix 00)—Screw: Rexolite, *sd* = 6.4 mm, *sl* = 20 mm)

Resonant Mode	Frequency (MHz)	Quality Factor, *Q*
1	271 (275)	1194 (1454)
2	757	1604
3	1183	1545

by about 8.5% and the Q_0 is reduced by approximately 3.8%. It can thus be concluded that, although the inclusion of such an element has a profound effect on the resonant frequency, its effect on the quality factor is moderate. The field distribution at the resonant frequency in the space inside the cavity with and without the tuning element is shown in Figures 4.16 and 4.17, respectively. As shown, the effect of the tuning element is to increase the electric field concentration inside the helix, thus contributing to an additional capacitive loading at the open end of the resonator. From Table 4.10, it is noted that the effect of the tuning screw on the resonant frequency is almost negligible in the presence of the dielectric screw, and therefore this technique is not particularly suitable for tuning in this case.

The effect of the screw diameter is demonstrated for this example in Figure 4.18(a), where the fundamental mode resonant frequency of metallic screws with fixed length, s_l, and varying diameter, s_d, is plotted against the latter. In the plot, the screw diameter, s_d, i_d normalized to the diameter of the helix, *d*. The frequency is normalized to the resonant frequency of the case without any tuning screw. Two values for the length of the screw are considered,—namely $sl_a = 22$ mm and $sl_b = 26$, that represent almost 28.75% and 41.39% of the axial length of the helix l, respectively. As shown, the effect of the screw diameter is substantial on the resonant frequency; thicker screws

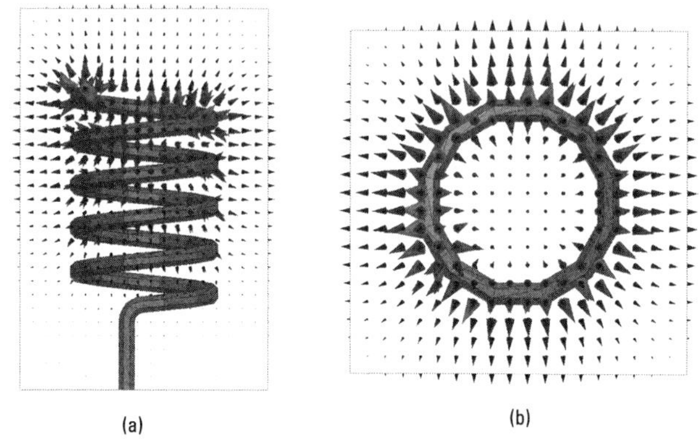

Figure 4.16 Helical resonator (helix00): Electric field distribution at the fundamental mode of resonance.

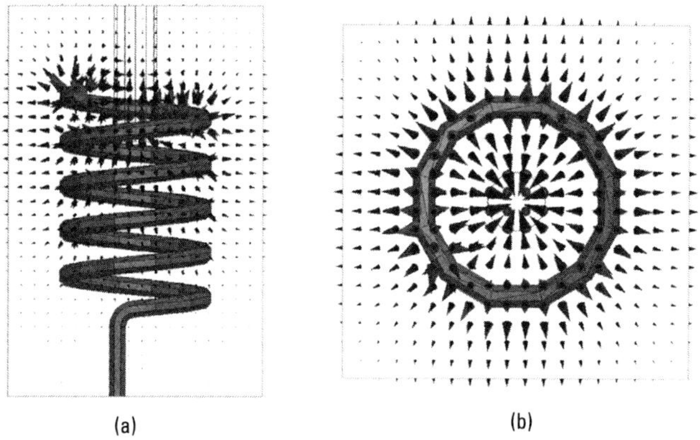

Figure 4.17 Metallic screw: electric field distribution at the fundamental mode of resonance (helix0–silver (Au) – s_d = 6.4 mm; s_l = 20 mm).

increase the capacitive loading to the helix and hence reduce the resonant frequency. An almost linear relationship is observed for both of the two screw lengths. The effect of the screw length is shown in Figure 4.18(b), where the normalized resonant frequency is plotted versus the length of the screw. The latter is normalized to the axial length of the helix. Two values of the screw diameter are considered, namely sd_a = 4.21 mm and sd_b = 8.42 mm, that present 20.0% and 40.0% of the helix diameter *d*, respectively. In this case, the effect in the resonant frequency of the inclusion of such a metallic screw

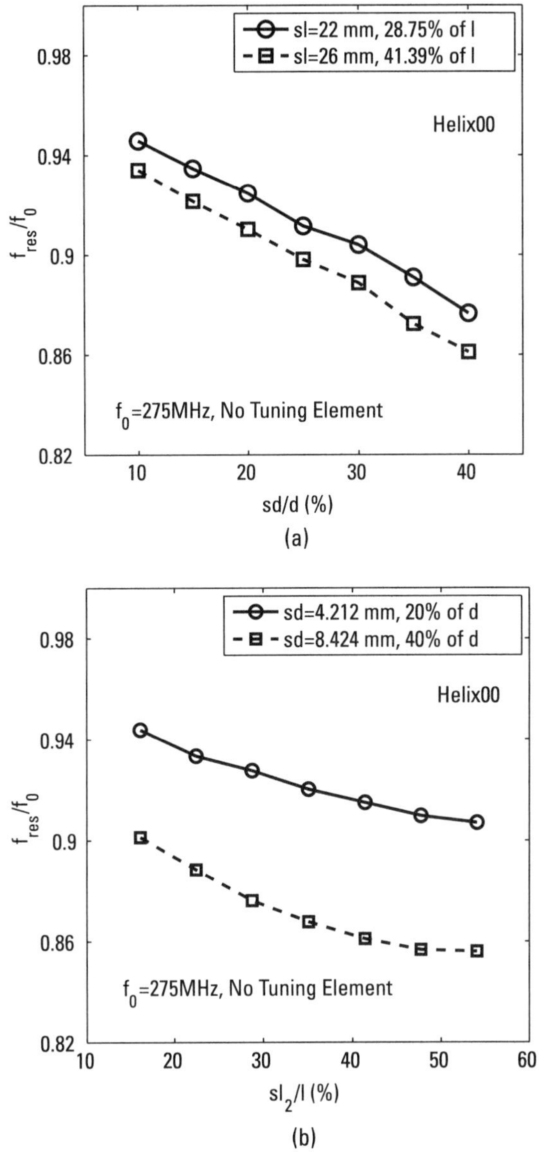

Figure 4.18 Effect of metallic (silver, Au) tuning screw penetration on the resonant frequency. (a) Fixed screw length, s_l, and (b) fixed screw diameter, s_d.

is less pronounced as compared to the case above. The above study suggests that thicker screws present more opportunities for resonant frequency tuning.

References

[1] Zverev, A., and H. Blinchikoff, "Realization of a filter with helical components," *IRE Transactions on Component Parts*, Vol. 8, No. 3, pp. 99–110, 1961.

[2] Zverev, A., Handbook of Filter Synthesis, New York: Wiley, 1967.

[3] Vizmuller, P., *Filters With Helical and Folded Helical Resonators*, Norwood, MA: Artech House, 1987.

[4] "ANSYS HFSS,"Ansys Inc., Available: http://www.ansys.com/Products/Simulation+Technology/Electromagnetics/High-Performance+Electronic+Design/ANSYS+HFSS.

[5] CST, Computer Simulation Technology, Darmstadt, Germany, 2005. Available at: http://www.cst.com/.

[6] Cameron, R., C. Kudsia, and R. Mansour, *Microwave Filters for Communication Systems: Fundamentals Design and Applications*, New Jersey: Wiley-Interscience, 2007.

[7] Pozar, D., *Microwave Engineering*, Third Edition, Hoboken, NJ: John Wiley & Sons, 2005.

[8] Sabbagh, M. E., K. A. Zaki, H. W. Yao, and M. Yu, "Full-wave analysis of coupling between combline resonators and its application to combline filters with canonical configurations," *IEEE Transactions on Microwave Theory and Techniques*, Vol. 49, No. 12, pp. 2384–2393, Dec. 2001.

[9] Young, L., "Q-factor of a Transmission Line Cavity," *IRE Transactions on Circuit Theory*, pp. 3–6, 1957.

[10] Wenzel, R., "Exact Theory of Interdigital Band-Pass Filters and Related Coupled Band-Pass Structures," *IEEE Transactions on Microwave Theory and Techniques*, Vol. 13, No. 5, pp. 559–575, 1965.

[11] Macalpine, W., and R. Schildknecht, "Coaxial resonators with helical inner conductor," *Proceedings of the IRE*, Vol. 47, No. 12, pp. 2099–2105, 1959.

[12] Doumanis, E., G. Goussetis, S. Kosmopoulos, and M. Guglielmi, "Conical-Shaped Helical Resonators for Improved Multipaction-Free Power Thresholds," *European Space Agency Microwave Technology and Techniques Workshop*, Nordwijk, 2012.

[13] Hunter, I., L. Billonet, B. Jarry, and P. Guillon, "Microwave filters-applications and technology," *IEEE Transactions on Microwave Theory and Techniques*, Vol. 50, No. 3, p. 794–805, 2002.

[14] Craven, G., and C. Radcliffe, "Microwave filters for communication satellites," *IEE Journal of Microwaves Optics and Acoustics*, Vol. 2, pp. 167–175, 1978.

[15] Fiedziuszko, S., "Satellites and microwaves," *14th International Conference on Microwaves, Radar and Wireless Communications (MICON)*, Gdansk, Poland, 2002.

[16] Boria, V., and B. Gimeno, "Waveguide filters for satellites," *IEEE Microwave Magazine*, Vol. 8, No. 5, p. 60–70, 2007.

[17] Russer, P., *Electromagnetics Microwave Circuit and Antenna Design for Communications Engineering*, Second Edition, Norwood, MA: Artech House Publishers, 2006.

[18] Jacoby, D., "Communication Satellites," *Proceedings of the IRE*, Vol. 48, No. 4, pp. 602–607, 1960.

5

Synthesis of Helical Resonator Filters

The design of microwave filters involves two main steps: the synthesis step (as briefly described in Chapter 3) and the physical implementation step. Once the prototype network (e.g., the coupling matrix) is computed, then the physical elements that can reproduce the same frequency response are calculated on a step-by-step basis. In Chapter 4, this process is described for individual helical resonators. This chapter describes the physical implementation of filters with helical resonators employing modern computer-aided design (CAD) tools. We employ a reference resonator design (as in Chapter 4) and demonstrate the technique for calculating the external coupling as applied to helical resonators and the technique to calculate the intercavity coupling utilizing a pair of helical resonators. A number of studies for the input/output and inter-resonator coupling are performed; these identify necessary know-how related to the specific technology. This way, we identify and highlight the peculiarities of the specific implementation technology and present solutions and methodology for an accurate and time-efficient design. Examples of a four-pole and a six-pole filter are employed to illustrate the methodology with a step-by-step approach. Finally, practical implementations of helical resonator filters and a diplexer are presented.

5.1 Design of Helical Resonator Filters

Prior to calculating the input/output (I/O) and inter-resonator coupling for helical resonator filters, it is necessary to have an accurate calculation of the resonant frequencies for individual helical resonators. This must be done within the filter structural layout that suits the filter implementation adopted according to the specification requirements (e.g., an inline filter or a folded filter with cross-couplings). The design time required to accurately calculate all the filter elements allows time-savings at the last stage of the final optimization, particularly for narrow band filters. Figure 5.1 shows example configurations of helical resonator models adopted to calculate the resonant frequency. To account for the iris loading effect, the neighboring physical space must be taken into account, as well as irises, tuning screws, and neighboring resonators. The resonators need to be detuned off resonance. This can be done by significantly altering the length of the resonators, by using auxiliary elements to ground the resonators, or by totally removing the resonators.

5.2 Input/Output Coupling

For helical resonator filters, coupling into and out of the filter can be achieved through a number of techniques [1–5] (e.g., through a wire loop (inductive loop), a disc probe (capacitive probe), or through tap coupling). For the wire loop coupling, a wire loop is positioned at the bottom end of the helical wire in a plane that is perpendicular to the axis of the helix. The coupling is controlled by the distance of the loop wire to the helical wire and by the diameter of the loop. This implementation can present manufacturing difficulties. For a disc probe coupling, a capacitive probe is positioned at one side of the cavity enclosure at the upper part of the helix (see the left side of Figure 5.2).

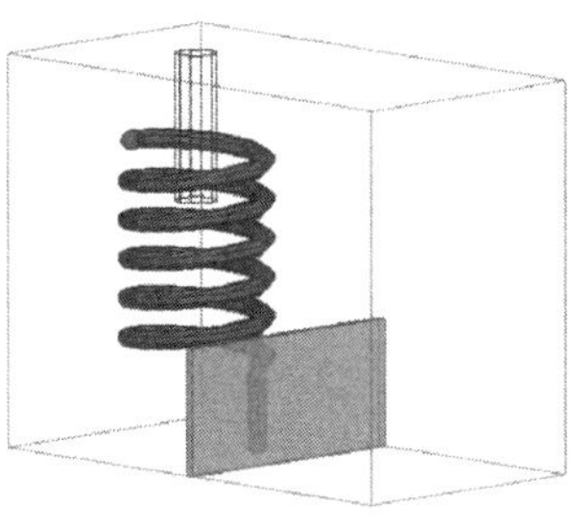

(a)

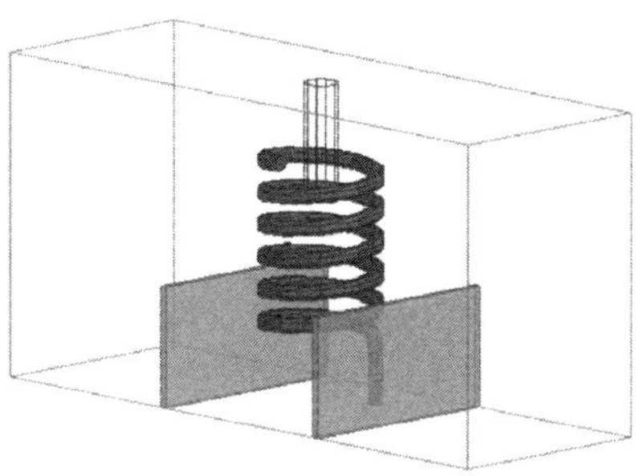

(b)

Figure 5.1 Model of a helical resonator within the metallic cavity enclosure employed for the accurate resonant frequency calculation for (a) an end resonator and (b) a middle resonator.

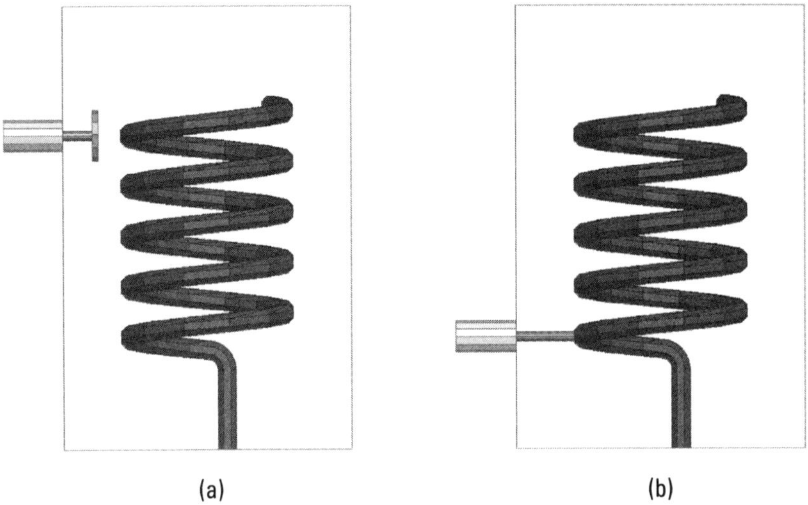

Figure 5.2 Input/output coupling implemented for helical resonators (a) disc probe and (b) tap coupling techniques.

For the tap coupling, a wire is connected to the helix wire at the lower end of the helix (grounded end) (see the right side of Figure 5.2). Special attention needs to be given to establish a good electrical and mechanical connection to the helix wire. The technique to achieve this connection is heavily dependent on the materials that are employed for the helical wire. Finally, rotation of the helix along its axis is required in order to establish the required input/output coupling.

Figure 5.2 illustrates example models of applied probe and tap coupling to helical resonators. The use of tap coupling has specific advantages in terms of improved mechanical stability under shock and vibration, easier realization, and is less prone to mechanical tolerances; it is therefore often used in practice. The loop and tap couplings are inductive couplings, whereas the probe coupling is mostly capacitive. In these final chapters, almost exclusively the tap coupling is considered.

In order to calculate the input/output coupling, the group delay method can be employed [6]. Alternatively, the frequency domain method can be used [6]. Both techniques are well-documented in [6, 7]. The application of the group delay method is based on the full-wave analysis of the reflection coefficient S_{11} looking into the input port of a single resonator. Post-processing allows the calculation of the reflection group delay from which the external quality factor, Q_e, and the normalized input impedance are obtained at the center frequency of the filter, f_0. The external quality factor Q_e and the normalized input impedance R_{Norm} of the coupling matrix model is given by ([6], Chapter 14).

$$Q_e = \frac{\omega_0}{R_{\text{Norm}} \bullet (\omega_2 - \omega_1)} \tag{5.1}$$

where $R_{\text{Norm}} = R/\omega_0 \cdot L \cdot FBW$, and $\omega_2 - \omega_1$ is the filter bandwidth in radians. The normalized input/output impedance of the coupling matrix model can be also obtained as [6]:

$$R_{\text{Norm}} = \frac{4}{\omega_2 - \omega_1} \frac{1}{\tau(\omega_0)} \tag{5.2}$$

where $\tau(\omega_0)$ is the maximum value of the group delay at the angular resonant frequency ω_0.

A model employed to apply the group-delay method is shown in Figure 5.2. A study that involved the rotation of the helix to achieve the required external coupling with and without a tuning screw was conducted. Figure 5.3 plots the effect of the tapping position on the group delay of the reflection

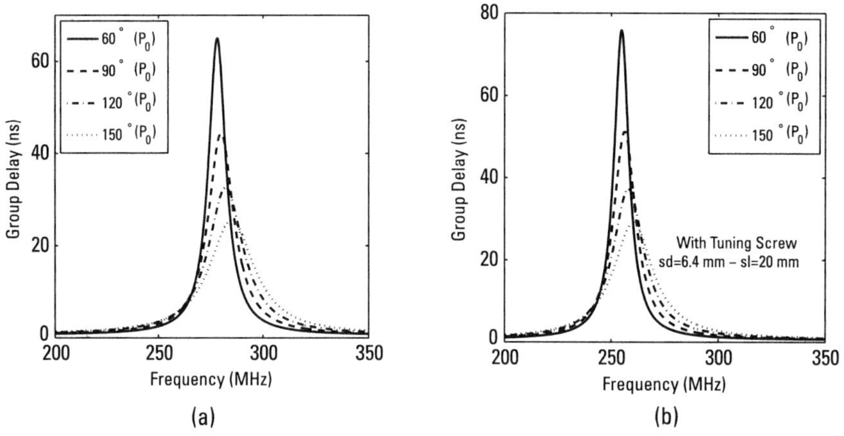

Figure 5.3 Group delay response for different tapping positions as a result of helix axial rotation (Figure 5.7); (helix00, Tuning screw: s_d = 6.4 mm, s_l = 20 mm) (a) without tuning screw, and (b) with tuning screw.

coefficient. The obtained results suggest a similar trend for both cases as the helix is rotated around its axis. It is noted that (as discussed in Chapter 3) the resonant frequency, indicated here by the group delay peaks, is shifted downwards in frequency with the tuning screw due to the capacitive loading. Figure 5.4 plots the extracted external quality factor, Q_e, obtained from the group delay analysis, as a function of the tapping position angle (see Figure 5.7). This information along with (5.1) can provide the normalized impedance R_{Norm} for the coupling matrix. Thus, by rotating the helix and calculating Q_e one can find the necessary tapping position to match the required R_{Norm} as provided by the coupling matrix synthesis. Alternatively, (5.2) and Figure 5.3 can provide the necessary tapping position. For a more accurate calculation of the I/O coupling, the iris loading effect should be accounted for in the simulation.

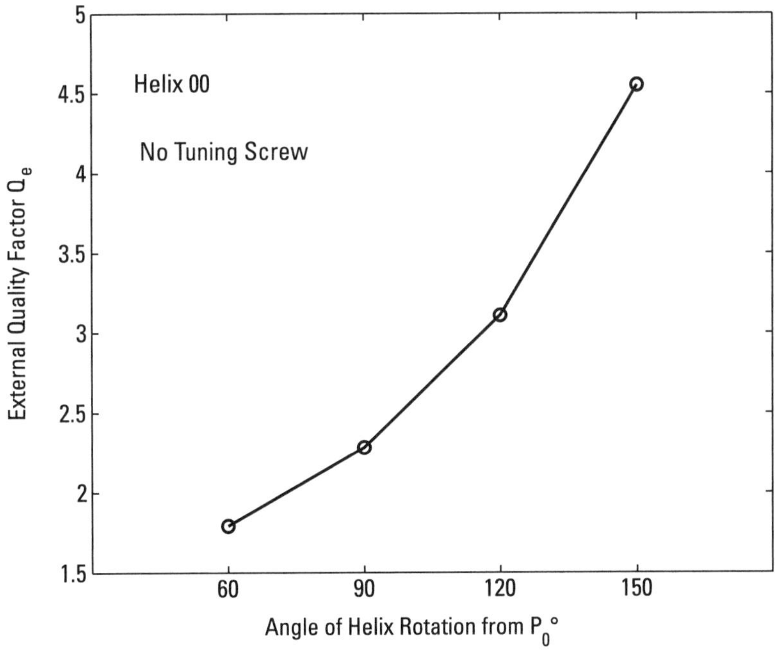

Figure 5.4 External quality factor Q_e as a function of tapping position (helix00, no screw). Figure 5.7 shows the model.

5.3 Inter-resonator Coupling

For the rigorous analytical calculation of the coupling, κ, of a pair of coupled resonators the equivalent lumped-element networks can be used. For the numerical computations of microwave resonators (distributed elements) the even- and odd-mode analysis can be applied [7]. This method is based on the observation that when two resonators synchronously tuned at f_0 in isolation (i.e., tuned to resonate at the same frequency, f_0) are electromagnetically coupled, then the natural resonance of the two-resonator system splits to two frequencies that correspond an even and an odd mode. Assuming for simplicity that the two-resonator structure maintains a geometrical symmetry along a given plane, the field distribution of the even/odd mode is such that it remains unaffected if a magnetic/electric[1] wall is placed along the symmetry plane. The coupling, κ, between the two resonators can then be obtained as [7]

$$\kappa = \frac{f_e^2 - f_m^2}{f_e^2 + f_m^2} \tag{5.3}$$

where f_m is the even mode and, f_e the odd mode resonant frequencies. The coupling coefficient M is given by:

$$M = \frac{f_0}{BW} \frac{f_e^2 - f_m^2}{f_e^2 + f_m^2} \tag{5.4}$$

where BW is the ripple bandwidth of the filter in MHz and f_0 is the center frequency of the pass-band filter. In general, $f_e > f_m$ for inductive and $f_m > f_e$ for capacitive coupling. In the case of asynchronously tuned resonators, formulas for the estimation of the coupling are given in [7].

In order to calculate the two resonant frequencies f_m and f_e, both an eigenmode analysis and an *S*-parameters analysis can be applied. It is noted that when geometrical symmetry exists, the computational problem can be reduced by applying electric and magnetic wall symmetries. When there is no geometrical symmetry, the entire structure is normally simulated (Figures 5.5 and 5.7) [7, 8].

For the eigenmode calculation, no excitation is required. The sign of the coupling, whether positive (inductive) or negative (capacitive), is determined

1. Electric/magnetic wall is used as the boundary condition where the tangential electric/magnetic field vanishes.

once the two resonant modes are identified by observing the field distribution at each mode. As an example of dominant capacitive coupling, the electric and magnetic field distributions of a coupled pair of synchronously tuned resonators are shown in Figures 5.5 and 5.6, respectively (Tables 4.1 and 4.2 provide the list of symbols for helices and cavity). Because there is no exact geometrical symmetry (due to the chirality of the helices), the entire coupled structure is analyzed, and the two natural resonant frequencies are identified.

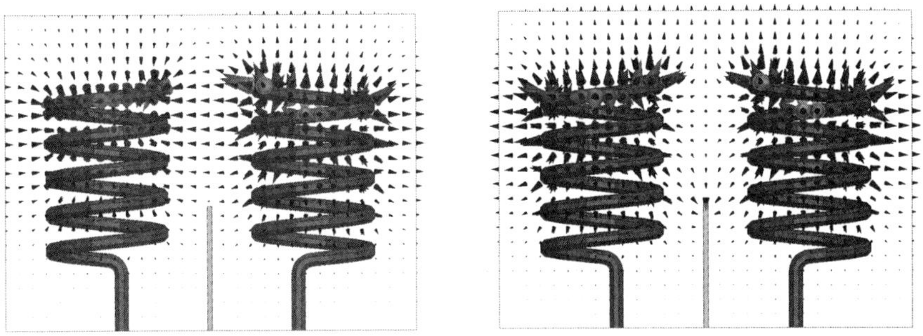

(a) (b)

Figure 5.5 Electric field intensity distribution of the two resonant modes for a pair of coupled helical resonators (helix00, *ih* = 22.6 mm, *it* = 0.9 mm). (a) Electric-type symmetry (identified as odd mode)—at low frequency, f_e. (b) Magnetic-type symmetry (identified as even mode)—at high frequency, f_m.

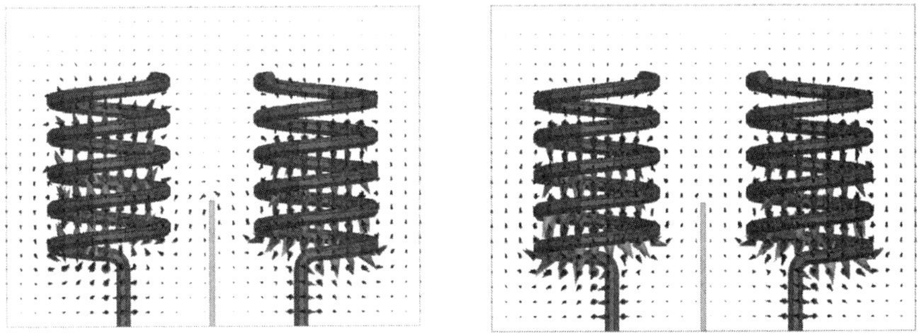

(a) (b)

Figure 5.6 Magnetic field intensity distribution of the two resonant modes for a pair of coupled helical resonators (helix00, *ih* = 22.6 mm, *it* = 0.9 mm). (a) Electric-type symmetry (identified as odd mode)—at low frequency, f_e. (b) Magnetic-type symmetry (identified as even mode)—at high frequency, f_m.

As shown in Figure 5.5, the fields at the low frequency respect electric-wall symmetry, and this frequency is identified as f_e. The higher frequency characterized by a magnetic-wall symmetry (magnetic fields note shown) is identified as f_m.

A model employed for the *S*-parameter calculation of the inter-resonator coupling is shown in Figure 5.7. This technique requires weak external excitation; the input ports provide weak excitation to the input and output resonators. In commercial software (e.g., high frequency structure simulator, or HFSS [9]), the weak coupling can be achieved by a post-processing step normalizing the impedance of the input/output ports to a high value [10]. To demonstrate the application of this technique, the same example as above has been considered. The magnitude and phase of the transmission coefficient S_{21} are plotted versus frequency in Figure 5.8. The two peaks of the magnitude represent the two natural frequencies [7, 8]. Typically, the wider the

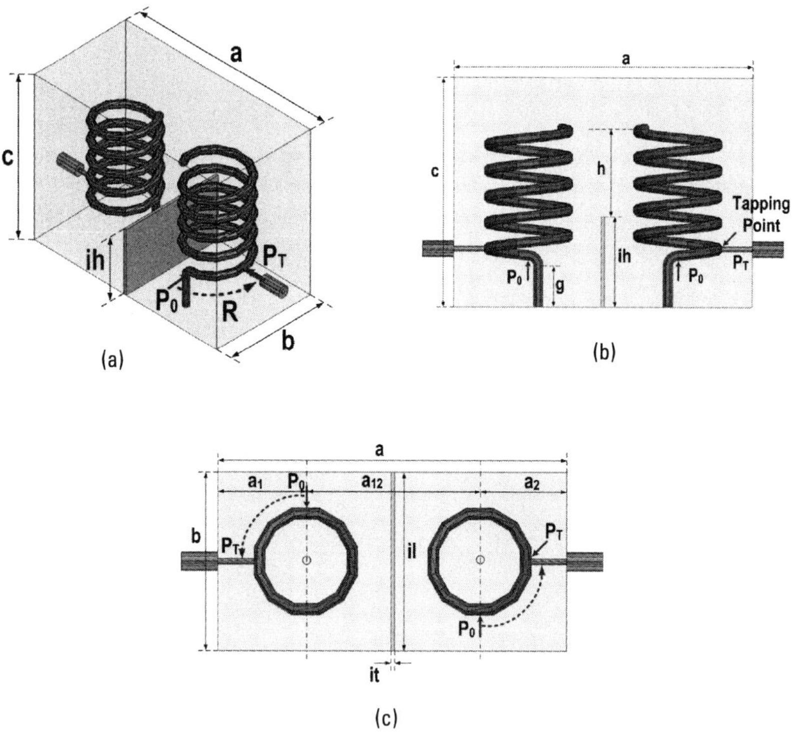

Figure 5.7 Structure configuration for the driven *S*-parameters calculation of the inter-resonator coupling (helix00, *it* = 0.9 mm).

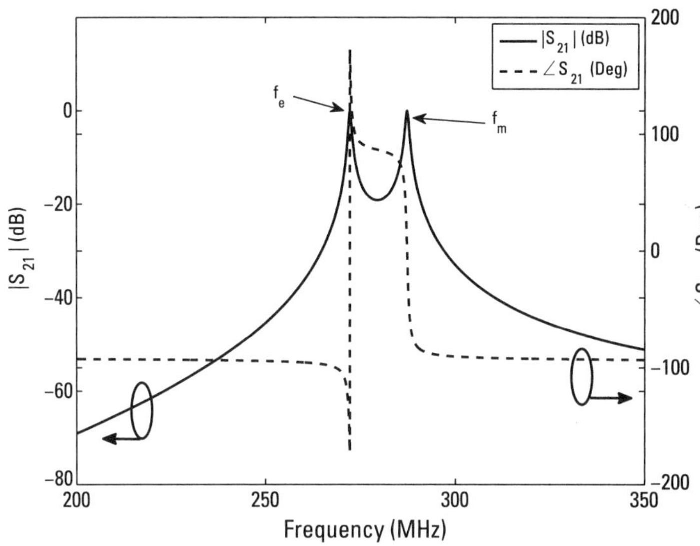

Figure 5.8 Typical *S*-parameters response over a frequency range around the center frequency (helix00, *ih* = 22.6 mm, *it* = 0.9 mm).

frequency separation of the two magnitude peaks, the stronger the coupling. To extract the sign of the coupling, one needs to compare the shift in the two natural frequencies with respect to the individual uncoupled resonant frequencies [7]. When comparing two coupled structures, an opposite phase response of the transmission coefficient, S_{21}, can be used as evidence of opposite sign of coupling [7]. A benefit of this technique is that it can be used to provide information on transmission zeros obtained as a result of resonant/mixed coupling [7, 10–14], as will be discussed in Chapter 6.

In the specific example considered here, the obtained values from the eigenmode analysis are f_e= 272.8 MHz and f_m=288 MHz, which give κ_{12} = –0.056, whereas from *S*-parameters analysis are f_e = 272.2 MHz and f_m = 287.3 MHz, which gives κ_{12} = –0.054. Both techniques, when properly applied, provide similar values for the coupling (also demonstrated later in this chapter, where a coupling study utilizing both techniques is reviewed).

Within a cavity enclosure, angular rotation of a helix along its longitudinal axis does not significantly alter the resonant frequency; however, the coupling between adjacent resonators when a helix is rotated is significantly affected. This is due to the fact that the electric field intensity distribution around the top open end of the helix is dependent on the helix ending position (see Chapter 4) and the field intensity distribution at resonance of standalone helical resonators. In view of Figures 5.5 and 5.7, the position of the helices is exactly the same in both cases. As discussed earlier, the tapping position dictates the position of the helix within the cavity. The effect of the tapping position to the calculation of the coupling is minimized when weak coupling is achieved.

5.3.1 Tuning Screws

Post-fabrication tuning of filters commonly requires tuning elements (screws). From a manufacturing point of view, tuning screws effectively address fabrication tolerance issues as they enable post-manufacturing adjustments. On the other hand, the inclusion of tuning screws increases the fabrication complexity of the filter and its physical footprint, while it may also give rise to other unwanted effects, like PIM [15]. The inclusion of the screws in helical resonator filters not only affects the resonant frequency of and unloaded Q-factor of the individual resonators (studied in Chapter 4), but also has a profound effect on the inter-resonator coupling.

To probe further into the adjacent resonator coupling, a study is conducted for a pair of resonators with and without a tuning screw coupled through an iris; in this case, the height of the iris is as a variable parameter. Figure 5.9 (a) shows a model employed to calculate the coupling. Figure 5.10 plots the coupling coefficient as a function of the iris height normalized to the axial length of the resonator. The coupling is capacitive and the common convention (negative) is used in this example. The results reveal that, as the height of the iris is increased, the coupling is decreased following an almost linear behavior for the cases considered. When the tuning screw is included, the coupling is significantly reduced, following a similar linear trend. This discrepancy between the two instances is attributed to the effect that tuning screws have on the electric field intensity distribution at the open end of the resonator. With the tuning screw, the fields are more confined. This decreases the coupling between adjacent resonators.

5.3.2 Decoupling Iris

Decoupling irises are used to control and suppress inter-resonator coupling in narrowband microwave filters. This section describes irises commonly employed in conventional helical resonator filters. It attempts to identify the coupling mechanism and provide guidelines for optimum design focused on

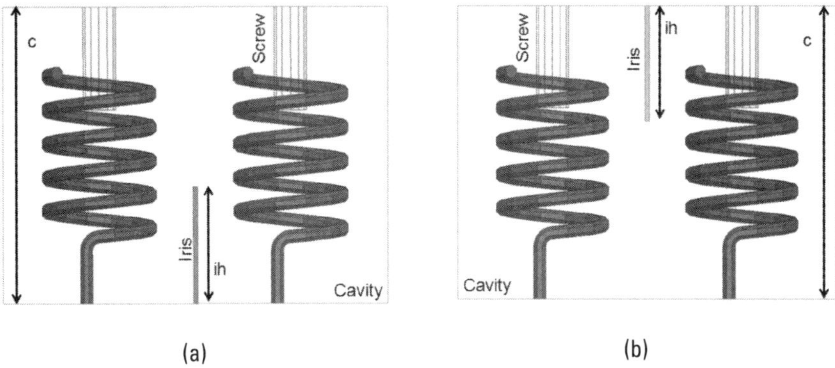

Figure 5.9 Model of a pair of helical resonators coupled with an iris. Parametric study with the iris height: (a) iris penetrating from the bottom of the cavity, and (b) iris penetrating from the top of the cavity.

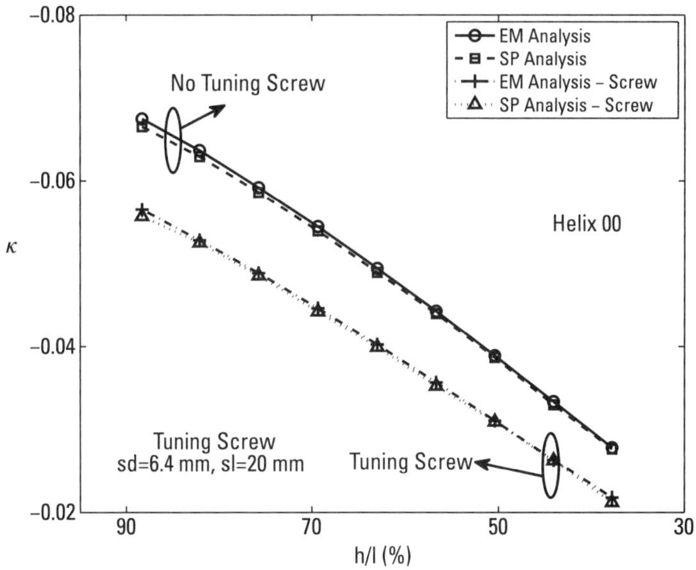

Figure 5.10 Coupling coefficient as a function of the height, *h*, normalized to the axial length of the resonator, *l*. Parametric study with the iris height, with and without a tuning screw (Figure 5.7—helix00, tuning screw: $sd = 6.4$ mm, $sl = 20$ mm, $it = 0.9$ mm).

physical implementation strategy. This is closely linked with the fabrication method of filter cavity enclosure (e.g., a milling process from the side of the cavity or from the top) and the tolerances (particularly since different types of irises can have smoother variation of the coupling coefficient with geometrical changes). In order to further assess the performance of coupling slots, two configurations are investigated:

- An iris wall penetrating from the top and bottom of the cavity;
- An iris penetrating from the two side walls of the cavity enclosure.

In the first case (Figure 5.9), an iris is penetrating from the bottom/top of the cavity enclosure at the side where the resonators are grounded/open-ended. Obtained results from the parametric study with iris length, *ih*, are summarized in Figure 5.12 (a). The plot shows the coupling κ as a function of iris height, *ih*, normalized to the cavity length, *c*. Within the majority range of iris length values, coupling is predominantly capacitive. Similar levels of maximum coupling values for both cases are reported when minimized values of the iris lengths are considered. For high values of the iris height, the coupling is reduced almost to zero in both cases. in the case of the top iris, there is a change in the sign of the coupling as the ih/c becomes 0.7 and more. This is attributed to the strong magnetic fields that are concentrated towards

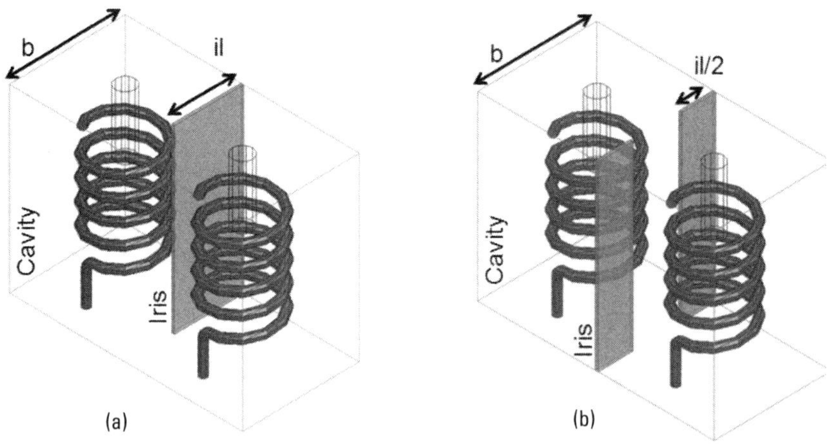

Figure 5.11 Model of a pair of helical resonators coupled with an iris. Parametric study of the iris height: (a) iris penetrating from one side of the cavity wall, and (b) iris penetrating from both sides of the cavity wall.

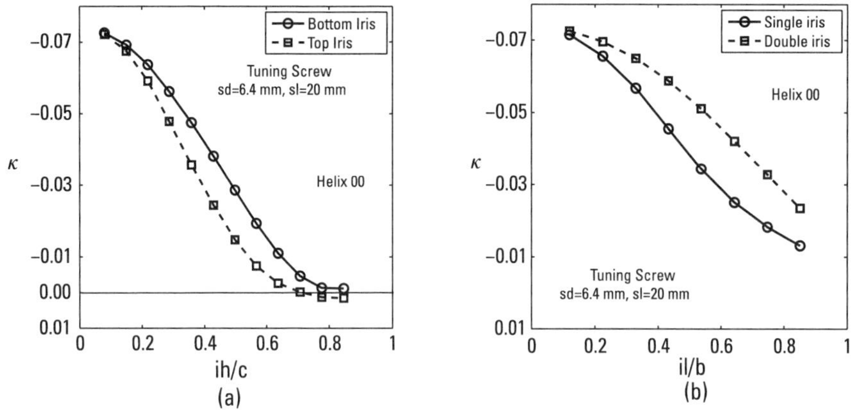

Figure 5.12 (a) Coupling, κ as a function of the iris height, *ih* normalized to the cavity height, *c*; parametric study with the iris height. (b) Coupling, κ as a function of the iris length, *il* normalized to the cavity width, *b*; parametric study with the iris length (helix00, tuning screw: $sd = 6.4$ mm, $sl = 20$ mm, $it = 0.9$ mm).

the shorted end of the helices and which predominantly contribute to the coupling in this case.

In the second case (Figure 5.11), an iris penetrates from the side walls of the cavity. The results are summarized in Figure 5.12 (b), which plots the coupling κ as a function of iris length, *il*, normalized to the cavity width, *b*. The obtained results demonstrate similar trends for the coupling. In this case, the coupling is predominantly capacitive within the entire range of iris lengths. Double iris demonstrates a smoother variation of coupling with the length of the iris. This iris can be chosen over the one considered above in the folded filter configuration due to its superior performance to suppress nonadjacent resonator undesired couplings.

5.3.3 Helix Axial Rotation

As mentioned earlier in this chapter, the coupling, κ, also depends on the relevant position of the resonators. This is further elaborated on in this section. Figure 5.13 shows two structural layouts of a model for a pair of coupled resonators. The study involves rotating the helix located at the left of the cavity enclosure, whereas the position of the helix located at the right is fixed. In the structure (a), the helix is not rotated (0 degrees), whereas in the structure (b), the helix is rotated by 180°. Figure 5.14 shows the inter-resonator coupling

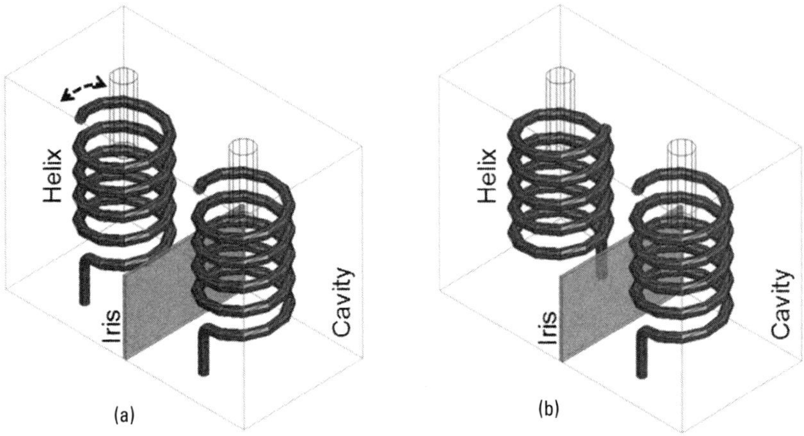

Figure 5.13 Structural layout of a model of a pair of helical resonators coupled with an iris: (a) the helix is fixed, and (b) the helix is rotated by 180°.

for a fixed iris height as a function of the helix angular rotation for a case with and a case without a tuning screw. The coupling coefficient is normalized to the obtained maximum value. The results demonstrate a discrepancy of the order of 10% for some of the rotation angles, which applies to both cases (with or without the tuning screw). It is also demonstrates the maximum coupling that occurs when two open ends of the helical wires are facing each other (similar to Figure 5.13 (a)). It is to be noted here that the magnitude of the coupling in the case with the tuning screws is reduced compared to the case without (as also reported above).

5.3.4 Effect of Cavity Width

In [16], the investigation of combline filters with increased spacing of the side walls ([16], Figure 1) is shown. The study considered the calculation of coupling as the side wall spacing (width of the cavity) was increased. A significant bandwidth and unloaded *Q* increase was shown for combline filters as the spacing-to-wavelength ratio was increased between 0.08 and 0.3 (wavelength at mid-band frequency of the filter). It is reported that the inter-resonator coupling in a general combline filter is due to evanescent waveguide modes (in the direction of propagation), and the TM_{11} and TE_{10} being dominant; these are similar to TEM modes (TEM static-field theory) when the ratio is small. As the side wall spacing increases, the waveguide modes become less

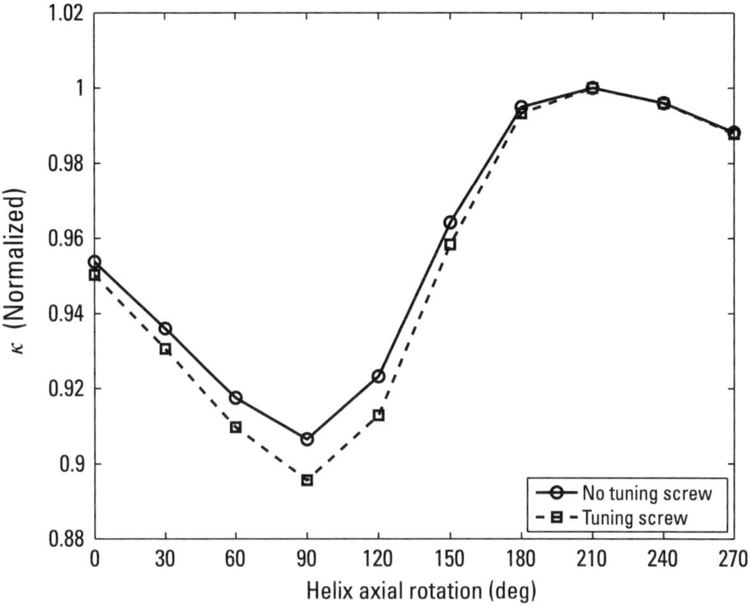

Figure 5.14 Coupling κ (normalized to maximum relevant value) as a function of helix axial rotation relevant angle (helix00, tuning screw: *sd* = 6.4 mm, *sl* = 20 mm, iris: *ih* = 22.6 mm, *it* = 0.9 mm).

attenuated (because the cut-off frequency of the waveguide is decreased), and the inter-resonator coupling increases [16]. The authors considered the filter as a transitional combline/evanescent mode structure.

Figure 5.15 reports a similar trend for the coupling between a pair of helical resonators. The study involves the calculation of the coupling as a function of the iris's height normalized to the axial length for three distinct values of the cavity width, b. The inter-resonator distance, a_{12}, was kept fixed in this study. Within the range of values considered for the height of the iris, the coupling is significantly increased as the cavity width, b, is increased.

5.4 Design Examples

Typically, the main electrical specification requirements involved in the design of band-pass filters include: desired center frequency, fractional bandwidth, maximum insertion loss, insertion loss variation in the passband, and several required rejection levels in the stopbands. This section demonstrates

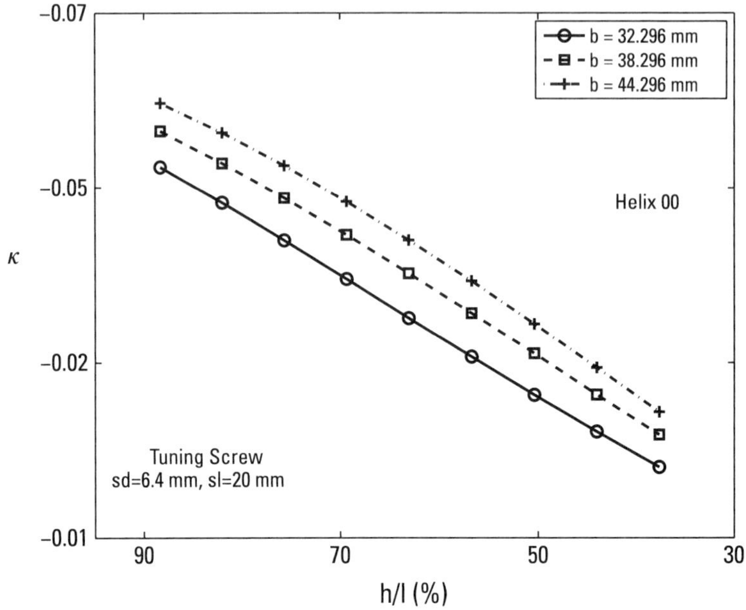

Figure 5.15 Coupling, κ, as a function of the height, *h* normalized to the axial length of the resonator, *l*; study of the cavity width *b* (helix00, tuning screw: *sd* = 6.4 mm, *sl* = 20 mm, *it* = 0.9 mm).

step-by-step the physical realization technique as reported in the two previous examples of helical resonator filters. As a first example, a fourth-order Chebyshev in-line helical resonator filter is designed according to the specifications. Subsequently, the design of a sixth-order Chebyshev filter in folded topology is presented.

5.4.1 Fourth-Order In-Line Filter

The specifications for the filter in this example are as follows: center frequency f_0 = 307 MHz, equiripple bandwidth, BW_{ripple} = 18 MHz (FBW = 5.86%), in-band return loss, RL = 21.9 dB (0.028 dB), rejection level > 50 dB for f= 260 MHz, and insertion loss variation < 0.05 dB. A Chebyshev transfer function is chosen over Butterworth due to its superior out-of-band rejection characteristics (see Chapter 3). The order of the filter is determined utilizing using [6, Chapter 3]. For Chebyshev filters, we only need to know the lowpass *g*-values in order to calculate the elements of the coupling matrix and/or the impedance and admittance inverter values. The obtained values from the synthesis are: $g_0 = 1$, $g_1 = 0.8569$, $g_2 = 1.2717$, $g_3 = 1.4934$, $g_4 = 0.7297$, and $g_5 = 1.1743$. The normalized impedances are, $R_{\text{Norm1}} = R_{\text{Norm4}} = 1.1670$, and the symmetric coupling matrix is shown below. The *S*-parameters response of the filter can be obtained using [6, Section 8.1]

$$M = \begin{bmatrix} 0 & -0.9759 & 0 & 0 \\ -0.9759 & 0 & -0.7256 & 0 \\ 0 & -0.7256 & 0 & -0.9759 \\ 0 & 0 & -0.9759 & 0 \end{bmatrix} \tag{5.5}$$

The physical dimensions for the helical resonators are derived with the synthesis procedure described in Chapter 4. The physical dimensions for the filter have been derived according to the synthesized values, the analysis for the input/output, and the adjacent resonator coupling. In Figure 5.16, a model of the optimized filter structure is shown. The initial simulated response of the filter is shown in Figure 5.17 (a). The superimposed prototype response is also shown for comparison. As can be seen, the initial obtained response matches the response of the prototype in terms of the center frequency and the bandwidth.

In order to further improve the filter's in-band return loss, further optimization EM-based techniques can be used [6, Chapter 15]. In the specific case, the HFSS optimizer (sequential nonlinear programming) has been used in two consecutive steps. Initially, the irises have been optimized to match to the filter bandwidth. Subsequently, the length of the tuning screws is optimized. The final response of the filter is shown in Figure 5.17 (b), where the simulated response and the coupling matrix response are in good agreement. The final dimensions of the filter are given in Tables 5.1 and 5.2. The simulated phase and group delay response of the optimized filter are given in Figure 5.18. The tuning method by Dishal [17] or the more synchronous group-delay tuning method of Ness [18] can be used for post-fabrication tuning.

With view of Figure 5.17(b) one can notice that although the in-band simulated response is in good agreement with the prototype response, this is not the case for the out-of-band response. The prototype, which assumes the coupling, κ, to be constant with frequency, exhibits an almost symmetric response around the center frequency where the rejection of the filter is increased monotonically. On the other hand, for the simulated filter, the out-of-band rejection is asymmetric and differs with frequency in the low and

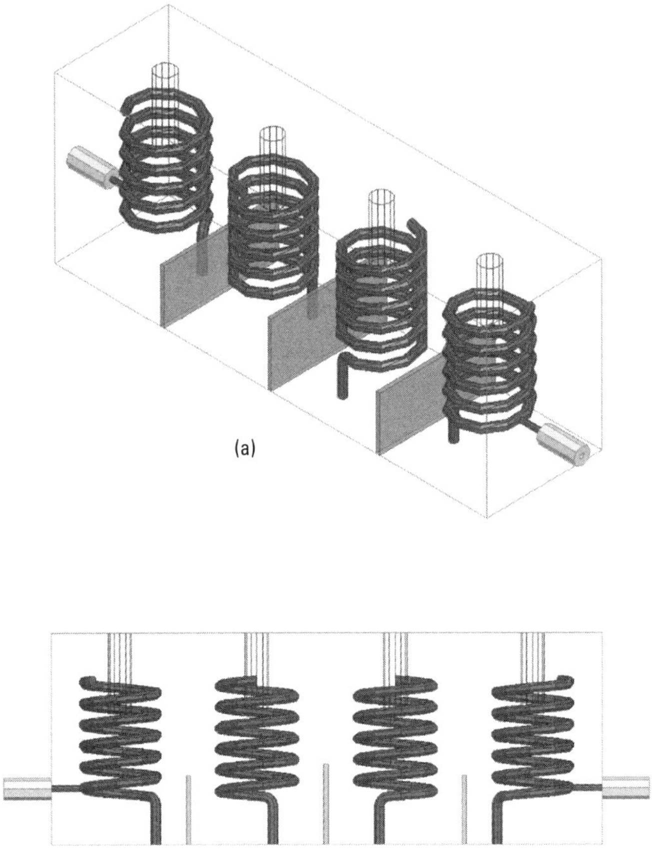

(a)

(b)

Figure 5.16 Model of a four-pole in-line filter.

high stop-bands. In particular, the simulated filter response has a steeper rejection at the low stopband, implying weaker couplings, as compared to the assumed frequency fixed by the protype. At the high stopband the rejection is degraded, indicating stronger coupling. The discrepancy can be attributed to a dispersive value of the coupling; this is very common in filter design and it is strongly related to the implementation technology (e.g., microstrip, waveguide, coaxial). The reader is referred to [19] for relevant information on coaxial cavity filters. It is further noted that the design approach presented above is valid for narrowband filters. The reader is referred to [20] for wideband and [11–14, 21] for mixed/highly dispersive coupling. More recently,

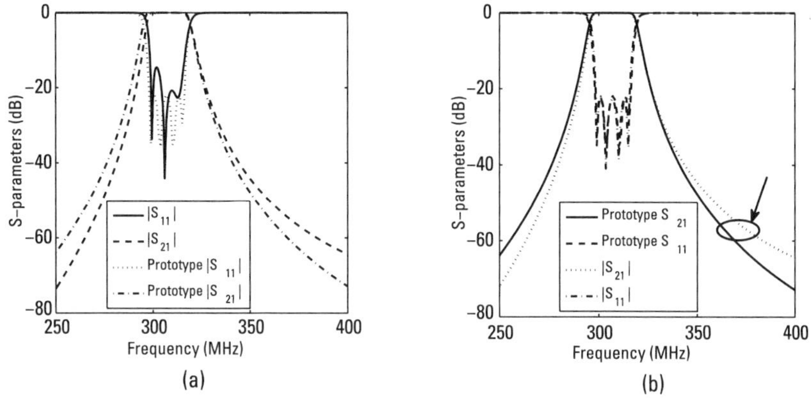

Figure 5.17 (a) Initial simulated response of the filter; prototype response superimposed for comparison. (b) Optimized simulated response of the filter; prototype response superimposed for comparison.

Table 5.1
Fourth-Order Helical Resonator Filter: Helix and Cavity Dimensions (Figure 5.16)

Helix	**mm(λ_0)**	**Cavity**	**mm(λ_0)**
d	15.75	$a_1=a_2$	14.52
p	4.28	b	30.64(.031)
l	23.66(.024)	c	42.95(.044)
w	2.05	$a_{12}=a_{23}=a_{34}$	28.64(.029)
n	5.53	$R_1=R_2$	148°
g	7.59		

a theory for frequency dispersive coupling matrix prototypes has been developed [20].

5.4.2 Sixth-Order Folded Filter

This section shows the physical layout and the obtained optimized response of a sixth-order filter in folded topology. This topology is commonly adopted for cross-coupled filters used to design quasi-elliptic filter functions or for more compact designs. In Figure 5.19 (a), the model of the sixth-order filter

Table 5.2
Fourth-Order Helical Resonator Filter: Dimensions (mm) of Screws and Irises (Figure 5.16)

Screw	1,4	2,3	Iris	1,4	2
sd	4.90	4.90	*il*	b	b
sl	16.06	15.51	*ih*	13.92	16.29
			it	0.9	0.9

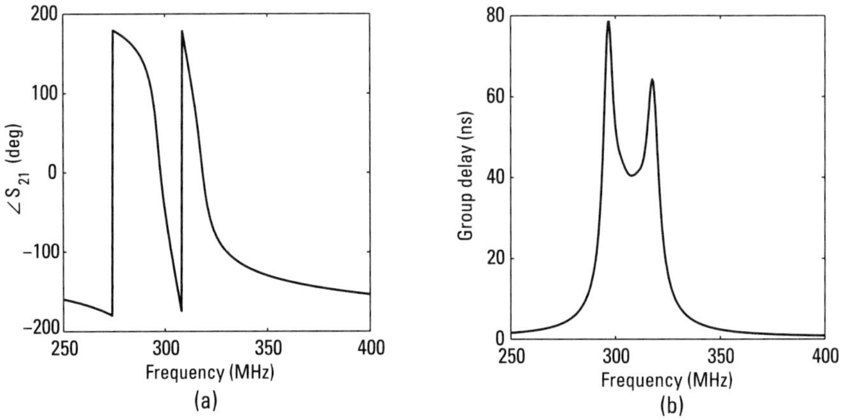

Figure 5.18 Optimized filter: (a) transmission phase and (b) group delay response.

is shown. The same helix is utilized in the fourth-order filter shown above. The simulated response of the filter after optimization is shown in Figure 5.19 (b). Due to the filter layout, undesired nonadjacent resonator coupling can exist. This unwanted coupling results in deviations of the filter response from the direct-coupled prototype response and creates problems in the optimization and post-fabrication tuning of the filter [19]. It is to be noted that the iris elements shown in Figure 5.11 would be beneficial in that perspective.

5.4.3 Realization Example

As an example of a real-life application of helical resonator filters, this section is dedicated to a brief description of the SRA diplexer, incorporated in the front-end of the payload ADCS-S&R embarked on the METOP satellite

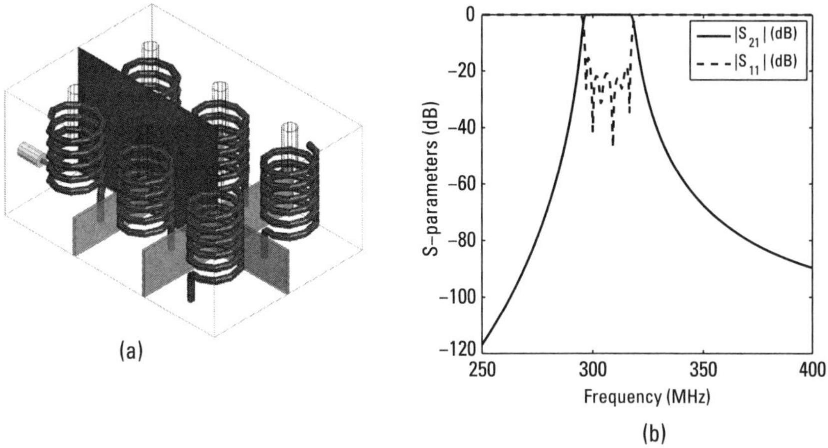

Figure 5.19 (a) Model of a sixth-order Chebyshev filter in folded topology, and (b) simulated response of the filter. No ohmic losses are assumed.

[22]. The SRA diplexer should satisfy typical electrical and mechanical requirements to be met in space applications throughout the lifetime of spacecraft, environmental conditions, interface, and measurements accuracy. The diplexer consists of an input section, two channel band-pass helical resonator filters and a number of low-pass filters at the chain of each branch of the diplexer. For channel 1 with the center frequency at f_0 = 243 MHz, a 0.01-dB bandpass ripple, and an equiripple bandwidth equal to 7 MHz. For the symmetric fourth-order Chebysheff filter utilized, the specification requirements consist of: a centre frequency at f_0 = 243 MHz, a 0.01 dB band-pass ripple, and an equiripple bandwidth equal to 10 MHz. The insertion loss allowed within the useful bandwidth, centered at the center frequency, should not exceed 1.2 dB for channel 1 and 1.1 dB for channel 2. A number of out-of-band rejection requirements are imposed for both channel filters. The return loss at the input and the output port of the unit at any frequency (within the operating frequency range) should not be less than 19 dB. The unit overall dimensions should be 350 × 140 × 120 mm (length × width × height) maximum. The mass of the equipment should not exceed 2.0 kg. Input and output connections should be TNC Coax (female). Temperature range of operation is –40°C to –65°C.

A photograph of the two channel filters is shown in Figure 5.20. Channel one filter is a symmetric, sixth-order Chebyshev filter. Input/output ports have been implemented through the tap coupling technique, as demonstrated

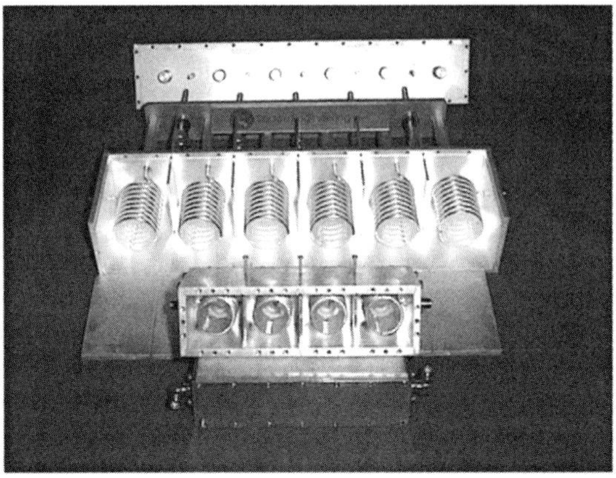

Figure 5.20 Two helical resonator channel filters for the METOP payload, designed and realized by the authors. (Courtesy Space Engineering S.p.A., Rome, Italy.)

earlier in this chapter. The dielectric support is chosen to be Rexolite 1422. This dielectric material is space qualified and is characterized by low losses. For this realization of the two channel filters, silver-plated aluminium has been used for the cavities, while the helices are made of silver-plated copper. All screws and the SMA connectors are commercial spares. SMA connectors have been chosen for the internal SRA diplexer interconnections, mostly due to their small size. The input and output taps have been realized using the core of the SMAs and have been fixed to the helix by stain solder. The three electric walls that separate adjacent resonators are not fixed (completely removable), and hence, the optimized definition of the dimensions of the coupling openings between adjacent resonators can be achieved. Figure 5.21 shows an open view of the flight model for the SRA diplexer. It clearly demonstrates the two channel helical resonator filters. Figure 5.22 shows the closed view of the diplexer and gives a sense of the overall size and mass.

5.5 Summary

This chapter is dedicated to showcasing the use of a step-by-step physical realization of helical resonator filters and utilizes two high-order filter examples. The elements of the coupling matrix are translated to physical dimensions of resonators, input/output coupling elements, and irises. Based on the studies for the external and inter-cavity coupling, the physical dimensions are obtained by matching the coupling matrix values. This is a useful and broad

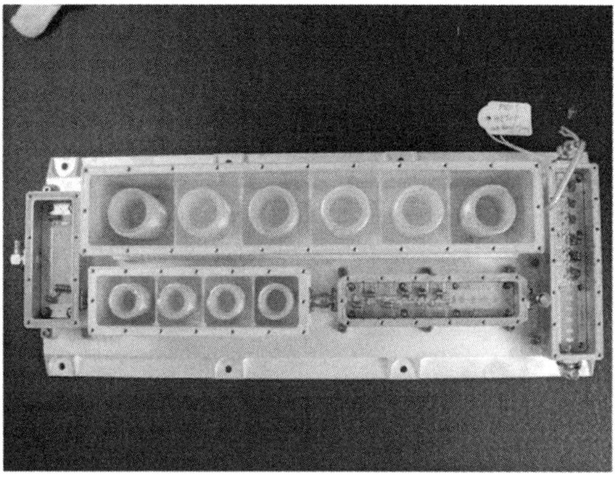

Figure 5.21 Flight model of the METOP SRA duplexer (open view), designed and realized by the authors. (Courtesy Space Engineering S.p.A., Rome, Italy.)

Figure 5.22 Flight model of the METOP SRA duplexer (closed view), designed and realized by the authors. (Courtesy Space Engineering S.p.A., Rome, Italy.)

approach that provides a good initial design for direct-coupled narrowband filters. Acceptable results can be obtained with minimum optimization effort.

In this chapter, the inter-resonator coupling is extensively investigated. A number of studies reveals that the coupling between high performance (high Q) helical resonators is mostly capacitive and is vastly independent on the type of irises employed. Furthermore, the coupling has proven dependent on the relevant position of the resonators. Subsequently, employing post-fabrication tuning screws significantly affects the coupling by reducing its magnitude.

References

[1] Macalpine, W., and R. Schildknecht, "Coaxial resonators with helical inner conductor," *Proceedings of the IRE*, Vol. 47, No. 12, pp. 2099–2105, 1959.

[2] Zverev, A., and H. Blinchikoff, "Realization of a filter with helical components," *Component Parts, IRE Transactions on*, Vol. 8, No. 3, pp. 99–110, 1961.

[3] Zverev, A., *Handbook of Filter Synthesis*, New York: Wiley, 1967.

[4] Vizmuller, P., "Filters with Helical and Folded Helical Resonators," Norwood, MA: Artecfh House, 1987, reference [3] in CH. 4.

[5] Tonne Software, Helical Software, Available: http://www.tonnesoftware.com/helical.html, 2010.

[6] Cameron, R., C. Kudsia, and R. Mansour, *Microwave Filters for Communication Systems: Fundamentals, Design and Applications*, Hoboken, NJ: Wiley-Interscience, 2007.

[7] Hong, J., and M. Lancaster, *Microstrip Filters for RF/Microwave Applications*, New York: Wiley-Interscience, 2001.

[8] Hong, J., and M. Lancaster, "Couplings of microstrip square open-loop resonators for cross-coupled planar microwave filters," *IEEE Transactions on Microwave Theory and Techniques*, Vol. 44, No. 11, pp. 2099–2109, 1996.

[9] HFSS, Ansoft, http://www.ansoft.com/products/hf/HFSS/. [Online]. Available: http://www.ansoft.com/products/hf/HFSS/.

[10] Wang, H., and Q. Chu, "An Inline Coaxial Quasi-Elliptic Filter With Controllable Mixed Electric and Magnetic Coupling," *IEEE Transactions on Microwave Theory and Techniques*, Vol. 57, No. 3, pp. 667–673, 2009.

[11] Hwang, H., and S. Yun, "The design of bandpass filters considering frequency dependence of inverters," *Microwave Journal*, Vol. 45, No. 9, pp. 154–163, 2002.

[12] Menzel, W., and M. Berry, "Quasi-lumped suspended stripline filters with adjustable transmission zeroes," in *IEEE MTT-S International Microwave Symposium Digest*, Vol. 3, 2004.

[13] Ma, K., J. Ma, K. Yeo, and M. Do, "A compact size coupling controllable filter with separate electric and magnetic coupling paths," *IEEE Transactions on Microwave Theory and Techniques*, Vol. 54, No. 3, p. 1113, 2006.

[14] Chu, Q., and H. Wang, "A compact open-loop filter with mixed electric and magnetic coupling," *IEEE Transactions on Microwave Theory and Techniques*, Vol. 56, No. 2, pp. 431–439, 2008.

[15] Macchiarella, G., G. B. Stracca, and L. Miglioli, "Experimental study of passive intermodulation in coaxial cavities for cellular base stations duplexers," *Proc. IEEE 34th European Microw. Conf.*, Amsterdam, The Netherlands, pp. 981–984, Oct. 2004.

[16] Levy, R., H. W. Yao, and K. A. Zaki, "Transitional combline/evanescent-mode microwave filters," *IEEE Transactions on Microwave Theory and Techniques*, Vol. 45, No. 12, pp. 2094–2099, 1997.

[17] Dishal, M., "Alignment and adjustment of synchronously tuned multiple resonant-circuit filters," *Proceedings of the IRE*, Vol. 39, No. 11, pp. 1448–1455, 1951.

[18] J. Ness, "A unified approach to the design, measurement, and tuning of coupled resonator filters," *IEEE Transactions on Microwave Theory and Techniques*, Vol. 46, No. 4, p. 343, 1998.

[19] El Sabbagh, M., K. Zaki, H. Yao, and M. Yu, "Full-wave analysis of coupling between combline resonators and its application to combline filters with canonical configurations," *IEEE Transactions on Microwave Theory and Techniques*, Vol. 49, No. 12, pp. 2384–2393, 2001.

[20] Meng, W., H.-M. Lee, K.A. Zaki, and A.E. Atia, "Synthesis of multi-coupled resonator filters with frequency-dependent couplings," *Proc. IEEE International Microw. Symposium Digest*, Anaheim, CA, 2010, pp. 1715–1719.

[21] Doumanis, E., G. Goussetis, and S. A. Kosmopoulos, "Inline Interdigital Pseudo-Elliptic Helical Resonator Filters," *IEEE Microwave and Wireless Components Letters*, Vol. 21, No. 8, pp. 400–402, Aug. 2011.

[22] Crino, V., and S. Kosmopoulos, "Metop—the SRA diplexer design-technical report," *Space Engineering S.p.A.*, Rome, Italy, RPT/METOP/00002/S, Tech. Rep., 1999.

6

Quasi-Elliptic Helical Resonator Filters

6.1 Introduction

In order to meet the increasingly stringent spectrum efficiency demands, modern communication systems require that filters have an improved skirt selectivity and stop-band attenuation. The out-of-band characteristics of filters can be effectively improved by the generation of transmission zeros in the stopband [1, 2] . As the name suggests, transmission zeros correspond to frequency points in the filter's transfer function where the transmission coefficient locally tends to zero (or negative infinite in dB). Employing filters with transmission zeros at finite frequencies is highly desirable because the advantage of their use includes: increased selectivity, sharp cut-off skirts, and asymmetric responses. It can allow to meet the out-of-band rejection requirements with smaller physical size and minimized insertion loss due to fewer number of resonators required [2]. Bandpass filters with finite transmission zeroes (often referred to as *quasi-elliptic filters*[1]) have been extensively considered in the literature [1–5]; their design and development remains a vibrant topic in the research community.

In this chapter, we focus on demonstrating techniques for introducing transmission zeros at finite frequencies in helical resonator filters. By means of examples, the techniques for developing pseudo-elliptic responses in helical resonator filters are illustrated. Two strategies are presented. A very widely used technique for introducing transmission zeros in the stopband exploits

1. In public literature, the term *quasi-elliptic* is often interchangeable with the term *pseudo-elliptic.*

coupling between nonadjacent resonators (cross-couplings). Section 6.2 presents example techniques for realizing cross-couplings in helical resonator filters. An alternative technique to implement transmission zeros at finite frequencies exploits in-line couplings that vary rapidly with frequency (highly dispersive). The application of this technique to helical resonator filters is demonstrated with examples in Section 6.3. It is noted that this chapter does not aim to provide a detailed treatment on synthesis techniques for filters with transmission zeros. The reader interested in the advanced filter design techniques is referred to the specialized literature referenced throughout this chapter.

6.2 Cross-Coupled Helical Resonator Filters

Cross-coupling of resonators is a widely employed technique to introduce additional TZs in band-pass filters [6–12]. It involves coupling (i.e., energy exchange) between nonadjacent resonators. When exploited to introduce transmission zeros at finite frequencies, the sign of the cross-coupling coefficient is typically opposite to that of the main path (in-line coupling) such that when the in-line coupling is electric, the cross-coupling is magnetic and vice versa. It is noted that cross-couplings are a generic strategy to adjust for the characteristics of the transfer function that can extend beyond the introduction of transmission zeros. For example, cross-couplings of the same sign as those along the main signal path can be exploited for group delay equalization purposes. The analysis and synthesis of cross-coupled resonator filter topologies have been extended to include the source and the load in addition to the resonators in the coupling matrix [8]; the interested reader is referred to the advanced filter synthesis literature [1–5] for more details.

When it comes to the physical realization, folded filter topologies may simplify the implementation of cross-couplings, as they can bring the cross-coupled resonators in closer physical proximity [2]. The presence of more than one cross-coupling may increase the realization complexity and create problems for the tuning of the filters [13, 14]. Moreover, as reported in [10], the folded filter topology can create unwanted cross-couplings that deteriorate the filter response and can prove problematic to the fine-tuning of the filter. Alternative cross-coupled topologies to simplify the physical realization involving cascaded trisections and quadruplets have been proposed [1, 15, 16], although these approaches give filters with a minimum number of transmission zeros. A very explanatory tutorial on the transmission zero mechanism of cross-coupled filters is presented in [17].

This section discusses the realization of cross-couplings in helical resonator filters. In particular, two cross techniques will be illustrated by means of filter realization examples namely: a) cross-coupling by a direct DC connection between nonadjacent resonators, and b) electromagnetic cross-coupling implementation in folded topology. This chapter does not extend beyond simple folded topologies suitable for a relatively small number of cross-couplings, although the implementation of other topologies can in principle follow the same strategies.

6.2.1 Cross-Coupling by Means of DC Connection

One technique for realizing cross-coupling between a pair of nonadjacent helical resonators involves the use of a wire that provides a DC electrical connection between the helices. This technique reduces any of the unwanted coupling effects mentioned above at the cost of increased manufacturing complexity. The technique is illustrated with an example reproduced from Yu and Dokas in [13] involving a seventh-order helical resonator filter centered at 306 MHz with a passband of 35 MHz and a transmission zero asymmetrically located at frequencies lower than the passband. The normalized input and output load impedances for this example are $R_{\text{Norm1}} = 1.1332$, $R_{\text{Norm7}} = 1.1774$, while the normalized coupling matrix is shown in (6.1)

$$M = \begin{bmatrix} 0.0123 & -0.8896 & 0 & 0 & 0 & 0 & 0 \\ -0.8896 & 0.0019 & -0.6150 & 0 & 0 & 0 & 0 \\ 0 & -0.6150 & -0.0607 & -0.5678 & -0.1103 & 0 & 0 \\ 0 & 0 & -0.5678 & 0.1147 & -0.5690 & 0 & 0 \\ 0 & 0 & -0.1103 & -0.5690 & -0.1027 & -0.6175 & 0 \\ 0 & 0 & 0 & 0 & -0.6175 & -0.0503 & -0.8985 \\ 0 & 0 & 0 & 0 & 0 & -0.8985 & -0.0075 \end{bmatrix} \tag{6.1}$$

As detailed in [2], the nonzero values of the diagonal elements represent the offset of each resonator from the filter's center frequency. As is further evident from this coupling matrix and the coupling diagram (6.1)—in addition to the couplings between resonators along the main signal path (i.e., resonator pairs (i, $i + 1$) with $i = 1, 2, \ldots, 6$)—a cross-coupling is introduced between the resonators (3, 5), as indicated by the nonzero values of the (3,5) and (5,3) of the coupling matrix. The cross-coupling is capacitive (negative) while the in-line couplings are all capacitive (negative). In order to enable the exchange of energy between the two nonadjacent resonators that will give rise to the desired cross-coupling, a metallic wire whose ends are attached on the helices of resonators 3 and 5 is exploited (in this example). By carefully selecting the position on each helix where the wire connects, it is possible to control both the sign of the coupling (i.e., whether the coupling is of magnetic or electric nature) and the value of the coupling coefficient. The analysis techniques described in Chapter 5 for the estimation of the coupling coefficient are also applicable in this case. It is noted that, when an EM simulator is used to obtain the coupling coefficient between two resonators in a multiresonator filter (see also Chapter 5), it may be useful to extend all other helices to electrically connect to the top wall. This is done in order to detune their resonance while still accounting for their presence in the electromagnetic environment of the pair under study.

Figure 6.1 shows a photograph of the prototype fabricated in [13]. A folded topology with two rows of three and four resonators (respectively) placed in parallel is used. Tap coupling to the input and output port is employed. By virtue of the folded topology, resonators 3 and 5 are brought physically in proximity, thus reducing the complexities associated with the

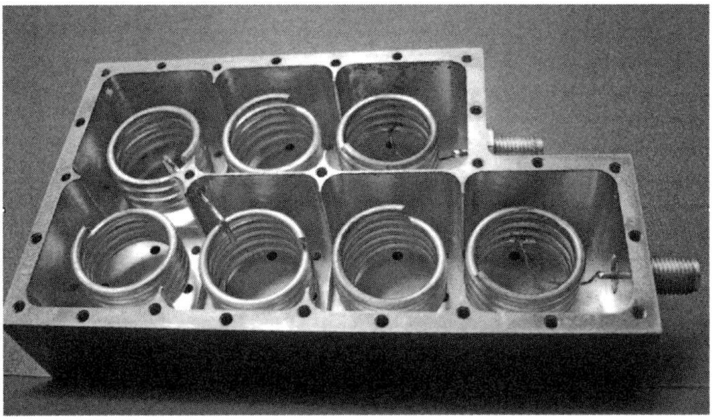

Figure 6.1 Seventh-order helical resonator filter with a cross-coupling between resonators 3 and 5. Here, cross-coupling is implemented by means of a wire that provides an electrically conductive connection between the two resonators. (The filter is reproduced from Yu and Dokas [13], 2014. Reprinted with permission.)

coupling wire. The latter can be seen penetrating through the metallic irises diagonally connecting resonators 3 and 5. Figure 6.2 shows the coupling diagram of the seventh-order filter. The simulated and measured response of this filter is shown in Figure 6.3. Excellent agreement between the two sets of results is observed. Seven reflection nulls can be seen in the passband commensurate with the order of the filter. Increased lower skirt selectivity is ob-

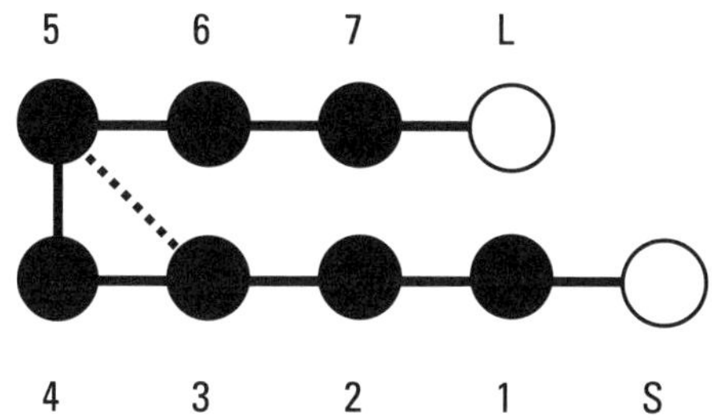

Figure 6.2 Coupling diagram of the seven-order filter in folded topology. Here, cross-coupling is shown between the third and the fifth resonators. Solid lines = main-line couplings, dashed lines = cross-couplings.)

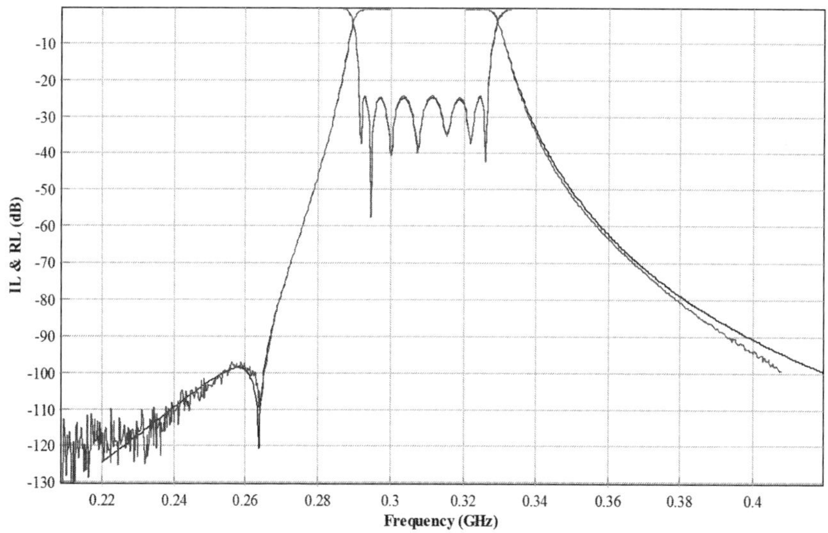

Figure 6.3 Simulated and measured results for the seventh-order cross-coupled filter shown in Figure 6.1. (The filter is reproduced from Yu and Dokas [13], 2014. Reprinted with permission.)

served as a result of the transmission zero at about 265 MHz. The measured Q-factor for the copper made filter was about 800.

6.2.2 Electromagnetic Cross-Coupling of Helical Resonators

An alternative cross-coupling realization relies on the exchange of energy through the electromagnetic fields between the nonadjacent resonators, much like the typical in-line coupling mechanism. In order to demonstrate an example of a fourth-order filter with electromagnetic cross-coupling for realizing transmission zeros in the stopband, the filter with response shown in Figure 5.17 is employed as a starting point. The aim is to implement the filter with coupling matrix shown in, but with additional positive cross coupling between resonators 1 and 4. In order to do so, a folded topology that brings resonators 1 and 4 in proximity is employed. A symmetric cross-coupling can then be achieved by using an iris between the first and last resonators of the filter.

Figure 6.4 shows a schematic layout of a fourth-order quasi-elliptic filter in a folded topology with electromagnetic cross-coupling. Resonators 1 and 4 are connected to the input and output ports, respectively. In-line couplings are realized by means of irises placed towards the open ends of the resonators, where the electric field is maximum, and therefore provides an electric (capacitive) coupling. In order to realize a magnetic (inductive) cross-coupling, the iris used between resonators 1 and 4 is towards the short-circuited end of

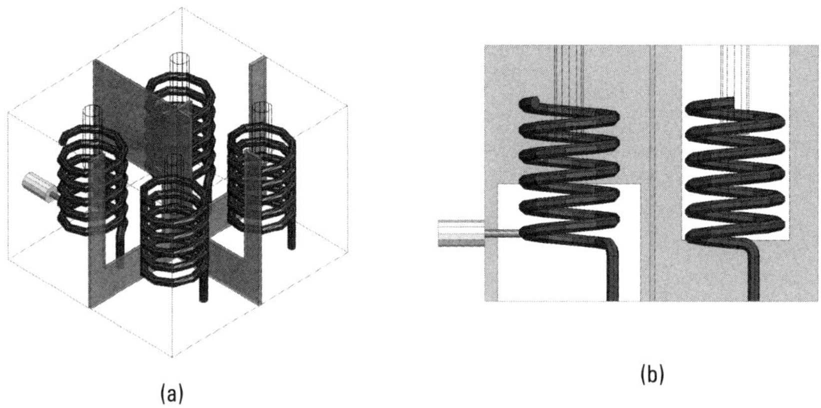

Figure 6.4 Fourth-order helical quasi-elliptic filter topology with electromagnetic cross-coupling.

the resonators where the magnetic field is stronger, and therefore the coupling is magnetic (inductive). A notable difference—as compared to the previous examples (Chapter 5)—is that the coupling irises are slightly altered to include vertical ribs (shown in Figure 6.4). The purpose of employing those is to reduce the unwanted coupling between nonadjacent resonators. The employment of these elements slightly reduces the overall coupling between the resonators, which can be increased by reducing the height of the iris.

Figure 6.5 plots the simulated response of the filter. The dimensions of the filter are given in Tables 6.1 and 6.2. The filter exhibits two attenuation poles placed symmetrically on the low and high stopbands as a result of the symmetric cross-coupling. The frequencies of the transmission zeros are at 260 MHz and at 357 MHz. Two transmission zeros are obtained in this case due to the symmetric cross-coupling [17]. The in-band return loss of the filter is of the order of 24 dB. Figure 6.5(b) plots the simulated magnitude of the transmission coefficient for four different heights of the iris. As the iris height is increased the frequency of the transmission zeros is moving closer to the passband, suggesting that the magnitude of the cross-coupling controls the frequency of the TZs. It is noted that in this examples the fourth order filter in folded structure topology are not equiripple, and further tuning is required.

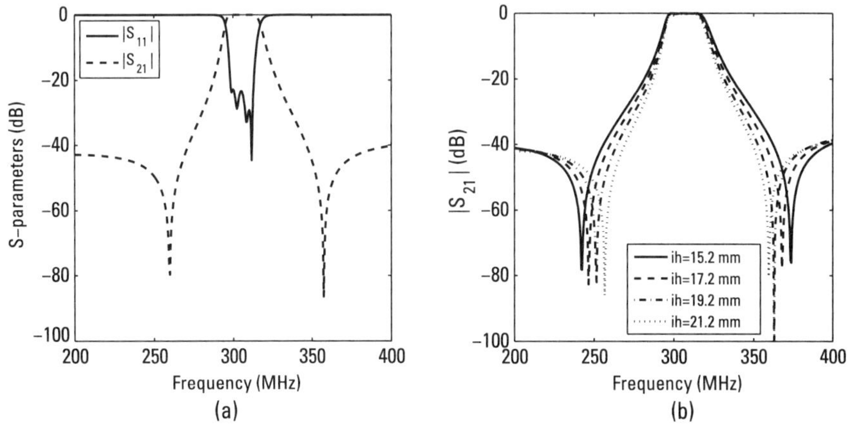

Figure 6.5 (a) Simulated response of the fourth-order quasi-elliptic filter of Figure 6.4. (b) Parametric study with the height of the iris demonstrating the ability to selectively locate two symmetric transmission zeros by altering the magnitude of the magnetic coupling.

Table 6.1
Fourth-Order Quasi-Elliptic Helical Resonator Filter:
Helix and Cavity Dimensions

Helix	mm(λ_0)	Cavity	mm(λ_0)
d	15.75	a_1=a_2	14.52
p	4.28	b	30.64(.031)
l	23.66(.024)	c	42.95(.044)
w	2.05	a_{12} =a_{34}	28.64(.029)
n	5.53	a_{23}	30.64(.031)
g	7.59	R_1= R_2	148°

Table 6.2
Fourth-Order Quasi-Elliptic Helical Resonator Filter:
Dimensions (mm) of Screws and Irises (Figure 6.5).

Screw	1,4	2,3	Iris	1,3	2	4
sd	4.90	4.90	il	b	b	b
sl	sl_1=16.12	sl_2=14.92	ih	9.26	10.23	25
	sl_4=15.70	sl_3=14.47	it	0.9	0.9	0.9

6.2.3 An Eighth-Order Example with Both EM and Wire-Enabled Cross-Coupling

The two approaches for implementing cross-couplings in helical resonator filters can be applied selectively. As an example that employs both electromagnetic cross-coupling and cross-coupling realized by a wire providing a DC path, an eighth-order filter reproduced from Yu and Dokas [13] is presented (see (6.2) and Figures 6.6–6.8). The filter is centered at 598 MHz with a bandwidth of 36 MHz. The coupling matrix of the prototype is given below:

$$M = \begin{bmatrix} 0.0207 & -0.8262 & 0 & 0 & 0 & 0 & 0 & -0.0166 \\ -0.8262 & 0.0196 & -0.5849 & 0 & 0 & 0 & 0.0679 & 0 \\ 0 & -0.5849 & 0.0206 & -0.5422 & 0 & 0 & 0 & 0 \\ 0 & 0. & -0.5422 & 0.0275 & -0.5817 & 0 & 0 & 0 \\ 0 & 0 & 0 & -0.5817 & 0.0026 & -0.5333 & 0 & 0 \\ 0 & 0 & 0 & 0 & -0.5333 & 0.0065 & -0.5608 & 0 \\ 0 & 0.0679 & 0 & 0 & 0 & -0.5608 & 0.0163 & -0.749 \\ -0.0166 & 0 & 0 & 0 & 0 & 0 & -0.749 & 0.0204 \end{bmatrix} \tag{6.2}$$

The normalized source and load impedances are $R_{\text{Norm1}} = 0.9622$ and $R_{\text{Norm8}} = 0.8262$, respectively. In this coupling matrix, cross-couplings exist between resonators pairs (1,8) and (2,7). The coupling diagram is shown in Figure 6.6. For the (1,8) pair, the cross-coupling is capacitive and therefore of the same nature as the in-line couplings; this cross-coupling contributes to additional transmission zeros. The cross-coupling between resonators (2,7) is inductive and gives rise to two additional symmetric transmission zeros. In terms of its implementation, a folded topology is employed. The photograph of the prototype is shown in Figure 6.7. The capacitive coupling between resonators (1, 8) is provided by means of an iris towards the open end of the resonators. The magnetic coupling between resonators (2, 7) is realized by means of a wire, as described in Section 6.2.1. The measured and simulated

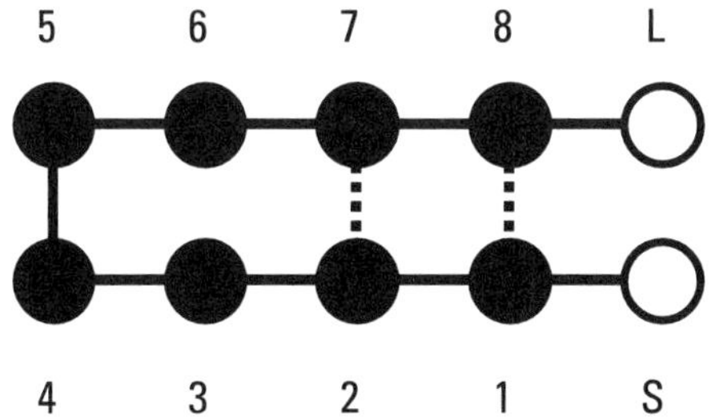

Figure 6.6 Coupling diagram of the seventh-order filter in folded topology. Here, cross-coupling is shown between the third and the fifth resonators (Solid lines = main-line couplings; Dashed lines = cross-couplings).

Figure 6.7 Eighth-order helical resonator filter with two cross-couplings between resonators (1,8) and (2,7). Here, cross-couplings are implemented by both electromagnetic fields (1,8) and by means of a wire that provides an electrically conductive connection between the two resonators (2,7). (The filter is reproduced from Yu and Dokas [13], 2014. Reprinted with permission.)

responses are shown in Figure 6.8. Excellent agreement is observed and alidates the circuit model. The measured Q-factor for the copper-made filter was about 1,100.

6.3 In-line Quasi-Elliptic Helical Resonator Filters

An in-line (rather than folded) resonator arrangement with finite transmission zeros can (in some cases) have comparative advantages when compared to cross-coupled arrangements. In-line arrangements can be preferred for the simplicity of their physical realization. Isolation problems can be avoided between input and output ports that may exist in filters that have folded configurations. Furthermore, the fact that in-line filters have their input and output ports at opposite sites allows easier arrangement in multiplexer and diplexer designs. In-line filters with finite frequency transmission zeros can be designed based on cross-couplings between multimode cavities [6] or extracted pole cavities [18]. An in-line topology for cross-coupled coaxial cavity filters is reported in [19].

An alternative technique introducing finite frequency transmission zeros in direct-coupled filters with in-line topologies is reported in [20–22]. In these designs, the coupling inverters exhibit a strong dispersion that provides exactly zero coupling at a given out-of-band frequency. An approximate synthesis approach that involves the prescription of finite frequency

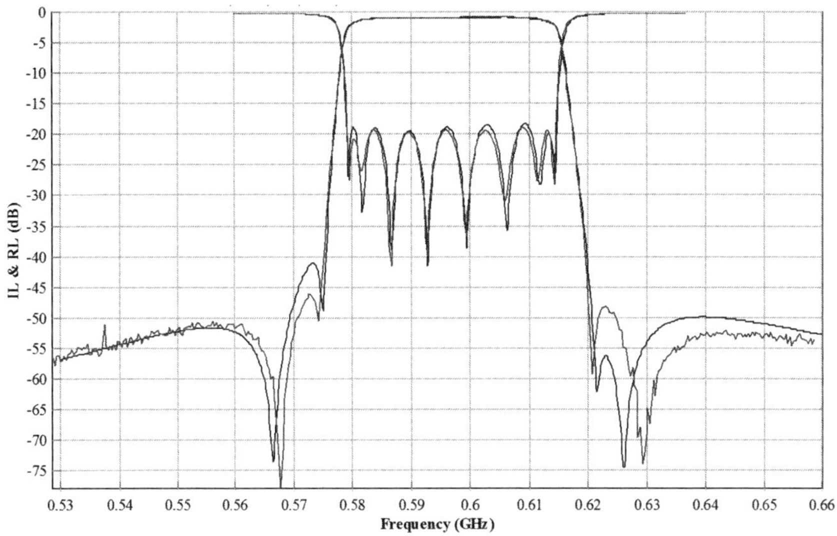

Figure 6.8 Simulated and measured results for the seventh-order cross-coupled filter shown in Figure 6.7. The filter is reproduced from Yu and Dokas [13], 2014. Reprinted with permission.)

attenuation poles to the inverters is given in [22]. For such designs, it is important to identify suitable geometries that provide the required highly frequency-dispersive behavior. In those cases, the high frequency dependence of the coupling allows a required value (as derived from the prototype) to be produced in the passband, while, at another out-of-band frequency, the coupling coefficient is zero, leading to the generation of a transmission zero. This technique (sometimes referred to in the literature as *mixed coupling*, due to the switch between inductive and capacitive coupling on either side of the transmission zero frequency) has been exploited to realize TZs in in-line topologies and various technologies in [23–26]. In [14, 26, 28] mixed cross-couplings are employed in order to realize additional transmission zeros.

It is noted that the strong frequency dispersion of the coupling employed in this class of pseudo-elliptic filters has been a problem in the design of direct coupled filters with moderate and wide bandwidths based on the classical theory [29]. The reason behind this lies with the fact that, when broadband responses are considered, all the basic elements of the filters (resonators and inverters) are frequency-dependent. As the filter relative bandwidth becomes wider, the circuit models with frequency-independent coupling do not accurately reproduce the response of the physical structure. The design of bandpass filters considering frequency-dependence of inverters has been reported in [30–32]. Frequency-dependent couplings have been explored in the synthesis of the coupling matrix network for the design of wide and ultrawideband filter structures [33]. However, a steep variation in the vicinity of the passband can be problematic, and as a result, mixed coupling concepts are challenging to implement for moderate and broadband filters, particularly when the TZ occur in the vicinity of the passband. Here, we present a technique for the design of narrowband in-line helical resonator filters with finite transmission zeros that exploits a dispersive coupling mechanism.

The dominant coupling mechanism in high power and high Q helical resonator filters in traditional combline topology is electric and is attributed to the dense electric field lines connecting the open ends of successive resonators [34]. The strong electric coupling overshadows some magnetic coupling that coexists in these configurations [34]. In practice, metallic irises are employed in order to control the value of the coupling coefficient in these filters [13, 34, 35]. In conventional helical filters, the inter-resonator coupling is mainly controlled by the height of the coupling iris. Assuming that the iris lies toward the open end of the resonator, it can be used to suppress magnetic coupling (and therefore increase the absolute value of the coupling coefficient), although smaller irises can eventually reduce the coupling coefficient.

In order to achieve strongly dispersive coupling, the helical resonators can be arranged in an interdigital topology so that their open ends are located at opposite sides. The layout for a second-order filter is shown in Figure 6.9. The coupling between the resonators in this configuration has both capacitive and inductive components (therefore termed "mixed" in [23–26]), each with a different frequency dependence (leading to a frequency-varying coupling and hence termed "dispersive" in [20–22, 30, 31]), and can be exploited to realize a TZ in the filter's stopband. The generation of the transmission zero is enabled by the different frequency-dispersion for the two coupling contributions (electric/capacitive and magnetic/inductive), which—at the frequency of the transmission zero—exactly cancel out while in the passband that is jointly provide the required level of coupling. With reference to Figure 6.9, it will next be shown that the coupling and the frequency of the transmission zero can be primarily controlled by three geometrical parameters, the helices overlap length L (linked with the distance to ground g), the cavity width b, and the inter-resonator distance a_{12}.

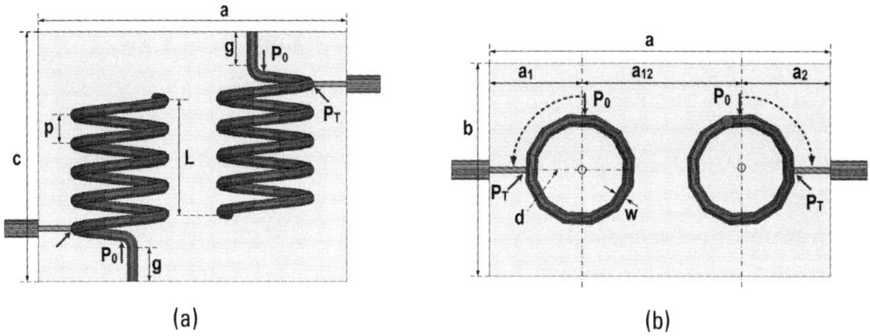

Figure 6.9 Topology of a pair of parallel coupled helical resonators in an interdigital configuration and physical dimensions.

6.3.1 Parallel-Coupled Helical Transmission Lines

Parallel-coupled transmission line filters [36, 37], are widely used in microwave circuits. The parallel coupling of resonators is shown to introduce TZs when different dispersion between even and odd modes occurs [38, 39]. Here, by means of an example involving two parallel-coupled helical transmission lines (as shown in Figure 6.10), it is shown that the parallel coupling can introduce a zero in the transmission of the coupled line structure. Full-wave electromagnetic modeling is employed. Each transmission line is excited by

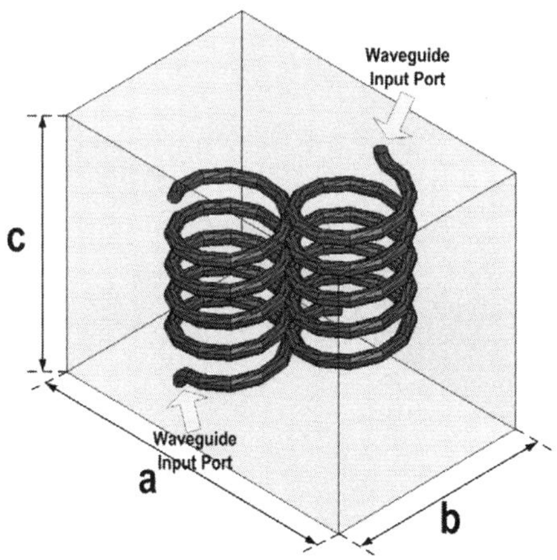

Figure 6.10 Topology of a pair of parallel-coupled helical transmission lines (in the case with zero ground distance *g*).

a coaxial waveport that is located at the point where the helix attaches to the cavity. The port emulates a semi-infinite transmission line of uniform cross section; if the impedance discontinuity between the helical coaxial transmission line and the port impedance is disregarded, then the structure depicted in Figure 6.10 comprises two semi-infinite transmission lines. With the exception of the relatively low reflection at the port interface, this arrangement therefore removes the strongly frequency-dependent effects associated with the quarter wavelength resonators otherwise encountered in filter configurations, thus enabling one to investigate in isolation the inherent dispersion of the interdigital coupling arrangement of the two helical coaxial lines. The dimensions of the helices and the cavity for the example that follows are given in Table 6.3. The case considered here is with the ground distance *g* zero.

Figure 6.11 shows parametric simulated results for the two-port transmission coefficient, which clearly demonstrates the emergence of a transmission zero. Given the aforementioned observation pertaining to the lack of any other strong resonant effects, this transmission zero is associated with a null coupling, which is in turn associated with this interdigital arrangement at a given frequency point. Investigation of the phase associated with this response reveals the strongly dispersive nature of the coupling mechanism. The change in the phase of the transmission coefficient from positive to negative values at the frequency of the transmission zero suggests that the coupling changes from predominantly capacitive to predominantly inductive. At the frequency of the transmission zero, the two contributions exactly cancel each other out and the transmission zero occurs.

According to the parametric studies in Figure 6.11, the frequency of the transmission zero is strongly dependent on the width of the cavity *b* and the overlap length *L*. It is noted that the overlap length *L* is varied by changing

Table 6.3
Dimensions of Helix and Cavity for the Parallel Coupled Helical Transmission Line Configuration

Helix	mm(λ_0)	Cavity	mm(λ_0)
d	21.06	a_1=a_2	19.15
p	6.53	a_{12}	30.3
l	31.63	b	Varied
w	2.61	c	Varied
n	4.85		

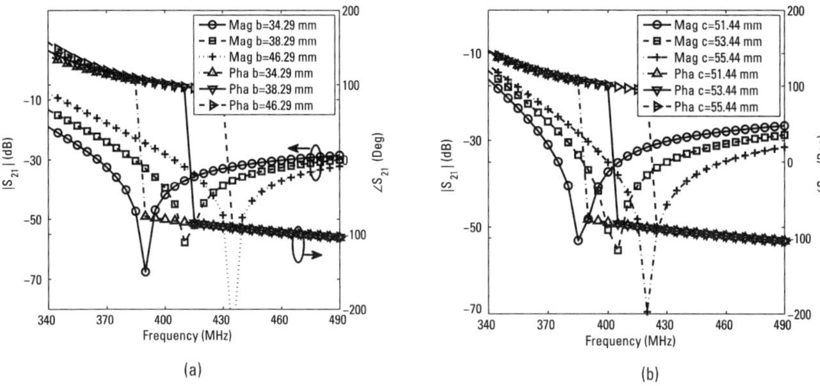

Figure 6.11 Magnitude and phase of transmission coefficient, S_{21} for the structure of Figure 6.10. Full-wave simulated results. Dimensions as in Table 6.3. Parametric study of (a) cavity width, and (b) overlapping distance.

the height of the cavity *c*. The obtained results in Figure 6.11 suggest that as the cavity width *b* is increased and the transmission zero is shifted towards higher frequencies. In a similar fashion, in Figure 6.11, as the cavity height, *c* is increased, which reduces the overlap length *L*, and the transmission zero shifts towards higher frequencies. Beyond providing an explanation of the nature of the transmission zero, the trends observed in these studies provide valuable design guidelines for the design of interdigitally coupled helical resonator filters.

6.3.2 Parallel-Coupled Helical Resonators

6.3.2.1 Mixed Coupling

The case of interdigitally coupled helical resonators is investigated next. Quarter wavelength transmission lines short- and open-circuited at either end are now employed, forming the two resonators that are otherwise excited using probe coupling (Figure 6.12). By means of two design examples (design 1 and design 2), the flexibility of the interdigitally coupled helical resonator arrangement used to deliver selectively electric or magnetic coupling of a given magnitude is present in the new configuration. The helix and cavity dimensions for both designs are given in Table 6.4.

According to the discussion in Chapter 5, the coupling is first investigated by means of a driven *S*-parameters modeling of the associated coupling

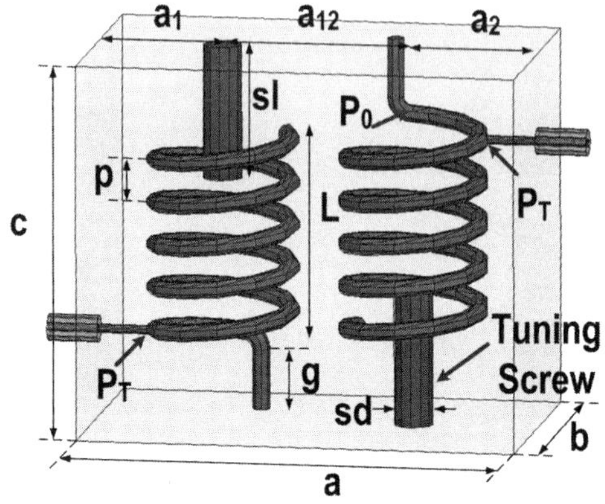

Figure 6.12 Layout of a second-order helical resonator filter in interdigital arrangement. (Copyright IEEE [40].)

Table 6.4
Dimensions of Helix and Cavity for the Two Designs, *e* and *m* (Figure 6.13)

Design 1				**Design 2**			
d	21.06	a_1	19.15	d	21.06	a_1	19.15
p	6.53	a_{12}	37.6	p	6.53	a_{12}	36.7
l	31.63	a_2	a_1	l	31.63	a_2	a_1
w	2.61	b	39.3	w	2.61	b	32.3
n	4.85	c	57.44	n	4.85	c	57.44
g	8.04	sd	5.0	g	13.39	sd	0
L	27.13	sl	19.75	L	2.43	sl	0

structure with weak external coupling. Using commercial software, weak input/output coupling is achieved by normalizing the port impedances to high values [26]. Figure 6.13 plots the magnitude and phase of the transmission coefficient S_{21} for the two designs. As shown, the split of frequencies associated with the even and odd resonances is identical across the two structures, indicating the same level of coupling. The opposite phase between the two designs implies that, in one case, the coupling is inductive (magnetic), and in the other, it is capacitive (electric).

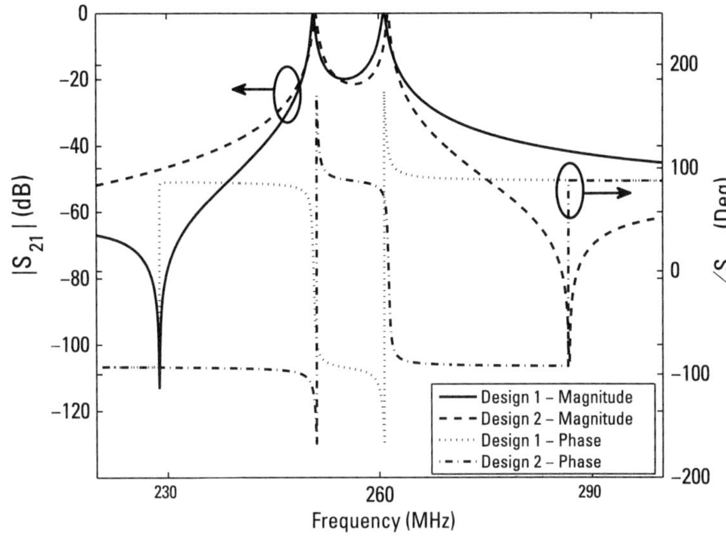

Figure 6.13 Transmission coefficients (magnitude and phase) for two pairs of synchronously tuned coupled helical resonators under weak external excitation. Dimensions as in Table 6.4. (Copyright 2014 IEEE [40].)

Eigenmode analysis can also be exploited in order to verify the nature of the coupling. The electric and magnetic field distribution of the two resonant modes for design 1 and design 2 is shown in Figure 6.14. For design 1, the resonance identified as odd resonance (which respects an electric-type symmetry [1]) occurs at lower frequency than the resonance identified as even (which respects a magnetic-type symmetry [1]), and reveals a predominantly electric coupling. On the contrary, the situation is reversed for design 2, revealing a predominantly magnetic coupling. The calculated values for design 2 are f_m = 265.23 MHz and f_e = 260.87 MHz, which give $|k_{12}|$ = 0.0166. For design 1, f_m = 260.37 MHz and f_e = 264.67 MHz, which give $|k_{12}|$ = 0.0164, in good agreement with the results of the weak coupling analysis. Because the even-type and odd-type resonant modes are interchanged in frequency for the two designs, this makes it clear that they have opposite signs for the coupling as has been shown from the phase response.

This analysis demonstrates that by varying the geometrical characteristics of the structure, it is possible to achieve predominantly capacitive or inductive coupling with a given magnitude. Furthermore, by exploiting the design flexibility of the additional parameters, one can achieve the same level of coupling with different frequency for the transmission zero and the same

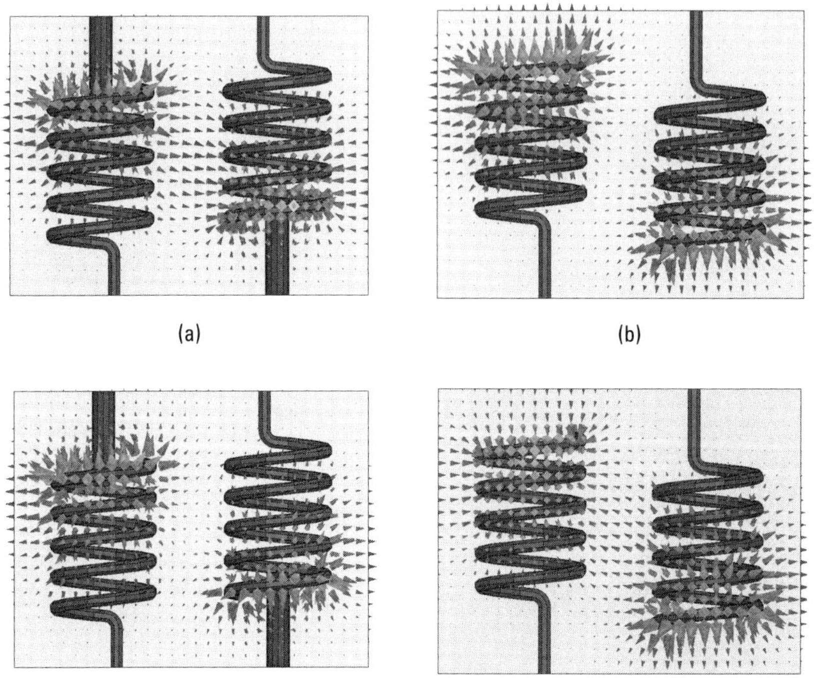

(a) (b)

(c) (d)

Figure 6.14 Electric field intensity spatial distribution of the two resonant modes for design E and design M. Identification for sign of the coupling. Dimensions as in Table 6.4. First resonant mode for (a) design E and, (b) design M. Bottom: Second resonant mode for (c) design E and, (d) design M.

transmission zero frequency for different levels of coupling. For a second-order mixed-coupled helical filter, the TZ will occur at the upper side of the passband when electric coupling dominates and vice versa when magnetic coupling dominates.

6.3.2.2 Dispersive Coupling

In order to demonstrate the frequency-dispersive nature of the coupling, a second-order filter is employed as an example. The topology of the filter is shown in Figure 6.9. The dimensions of the helices and the cavity are given in Table 6.5. Figure 6.16 (a) plots the simulated filter response as obtained from the full-wave tools. The center frequency of the filter is 291 MHz and the ripple bandwidth is 11 MHz, which suggests a fractional bandwidth of 3.8%. The filter exhibits an out-of-band TZ above the passband approxi-

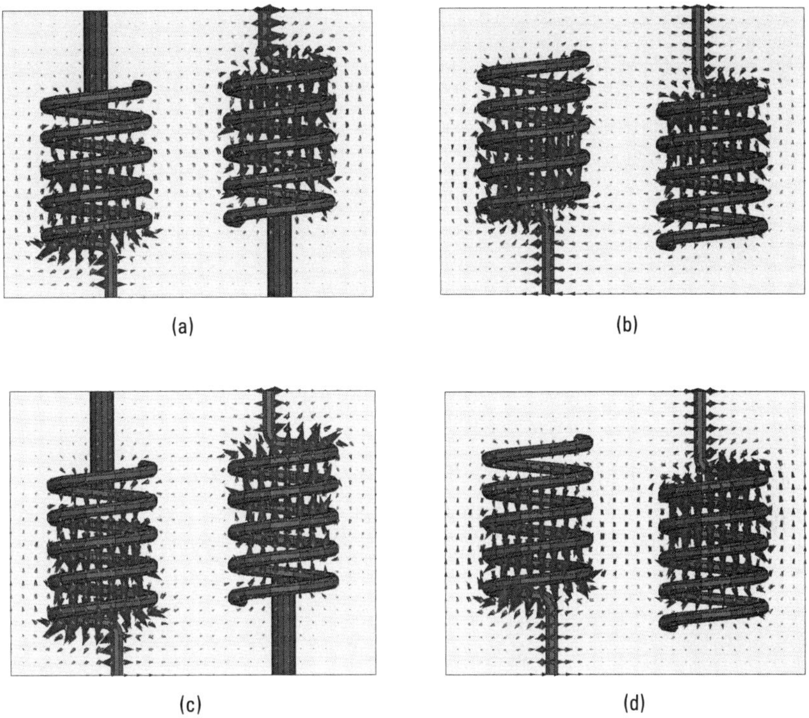

Figure 6.15 Magnetic field intensity spatial distribution of the two resonant modes for design E and design M. Identification for sign of the coupling. Dimensions as in Table 6.4. Parametric study of (left) cavity width, (right) overlapping distance. First resonant mode for (a) design E and, (b) design M. Second resonant mode for (c) design E and, (d) design M.

Table 6.5
Dimensions of the Second-Order Helical Filter; Helix and Cavity

Helix	mm(λ_0)	Cavity	mm(λ_0)
d	21.06	a_1=a_2	19.15
p	6.53	a_{12}	34.3
l	31.63	b	44.3
w	2.61	c	57.44
n	4.85		
g	8.04		

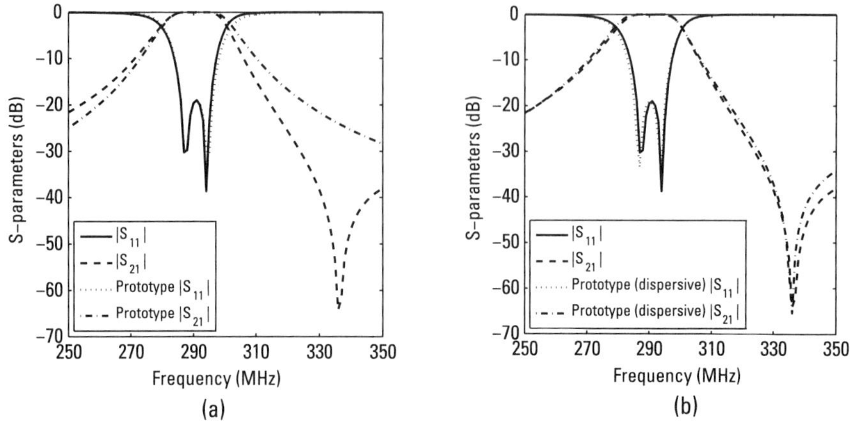

Figure 6.16 Second-order filter design. (a) coupling matrix prototype and simulated response of the filter. (b) Coupling matrix prototype (imposed linear frequency-dispersive coupling) and simulated response of the filter. Dimensions of the second-order filter are as in Table 6.5.

mately at 336 MHz. Figure 6.16 (b) also plots the response of a coupling matrix prototype for a direct-coupled filter with the same center frequency and bandwidth. The obtained values from the synthesis for the g_k parameters are: $g_0 = 1$, $g_1 = 0.7251$, $g_2 = 0.5742$, $g_3 = 1.2626$. The normalized impedances are $R_1 = R_4 = 1.4228$, and the 2×2 symmetric coupling matrix is

$$M = \begin{bmatrix} 0 & -1.5806 \\ -1.5806 & 0 \end{bmatrix} \tag{6.3}$$

The coupling coefficient at the center frequency has been calculated with eigenmode analysis. Given the coupling matrix $[M]$, the filter's S-parameters can be obtained. The response of the simulated filter matches well the response of the prototype in the frequency range close to the filter's passband.

In order to apply a first approximation of the coupling as a function of frequency, we commence from calculating the coupling coefficient at the center frequency. We next impose that the coupling at the frequency of the transmission zero should be zero. Based on these two points, a linear approximation for the coupling coefficient with respect to frequency is derived for the frequency range from the center frequency to higher frequencies.

Assuming that the coupling coefficient is constant at all frequencies below the passband, an approximation for the coupling coefficient is derived for all the required frequencies both below and above the passband of the filter. The next step is to include this information into the coupling matrix (assuming frequency-dependent coupling coefficients). The prototype response with frequency-dependent coupling coefficients is shown in the right half of Figure 6.16. It can be seen that the response (as obtained from the coupling matrix prototype) matches the simulated filter response.

Derivation of the Coupling Coefficient

In the analysis that follows, the concept of impedance/admittance inverters is employed [3]. We employ a second-order filter as an example to estimate the coupling coefficient. This estimation of the coupling coefficient is based on the insertion loss that is produced by the discontinuity and the reactance slope parameters x of the resonators [3]. The discontinuity between the two parallel coupled helical transmission lines can perform as an impedance inverter. This is attributed to the general property of lossless passive reciprocal networks; accordingly, any such network can be converted to an equivalent K-inverter [41]. It is possible to choose reference planes at electrical distances -1 and -2 from the respective ports in order to convert the asymmetric two-port network into a symmetric impedance inverter [41]. These reference-plane locations are given in [41].

The configuration employed for the full-wave calculation of the K-inverter value is shown in Figure 6.17. The input impedance at the ports must be matched with the use of the full-wave software to the characteristic impedance of the helical transmission lines Z_0. The overall two-port network is a K-inverter and the impedance of the K-inverter, K can be calculated from the S-matrix of the structure according to:

$$K = \sqrt{\frac{Z_{01}}{Z_{02}}}\sqrt{\frac{1-|S_{11}|}{1+|S_{11}|}} \tag{6.4}$$

where Z_{01} and Z_{02} are the characteristic impedances of the transmission lines. The dimensions for the cavity and the helices are exactly the same as the dimensions of the second-order filter and are given in Table 6.5. In order to derive the coupling coefficient, we start from a frequency-dependent K-inverter. A full-wave simulator is used to calculate the S-matrix (which, of

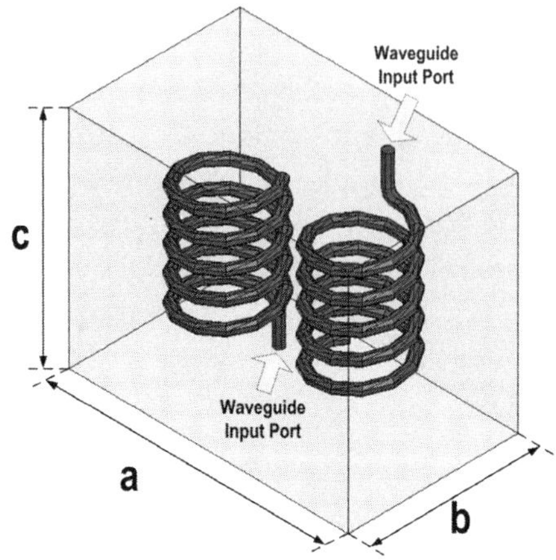

Figure 6.17 Topology of a pair of parallel coupled helical transmission lines.

course, includes the frequency-dependent information for the discontinuity). The amplitude of the reference scattering parameters $|S_{21}|$ and $|S_{11}|$ are derived. This information is included in the calculation of the coupling coefficient as follows.

For a resonator exhibiting a series-type resonance (zero impedance at the resonant frequency), the reactance slope parameter is given by [3]:

$$x = \frac{\omega_0}{2} \frac{dX}{d\omega}\bigg|_{\omega=\omega_0} \tag{6.5}$$

where ω is the angular frequency, and $X(\omega)$ is the reactance of the distributed resonator. The electrical length of the resonator θ can be expressed as:

$$\theta = \frac{2\pi}{\lambda_g} L_{res} - \phi \tag{6.6}$$

where λ_g is the guided wavelength, *Lres* is the physical axial length of the resonator and ϕ is a constant that accounts for the electrical length reduction due to the coupling. Then x can be expressed as:

$$x = Z\frac{\omega_0}{2}\frac{d\theta}{d\omega}_{|\omega=\omega_0} \tag{6.7}$$

where Z is the characteristic impedance of the transmission line. For a homogeneous transmission line, the reactance slope parameter is given by ([3] Chapter 5),

$$x = \frac{(2n-1)}{4}\pi Z_0\left(\frac{\lambda_{g0}}{\lambda_0}\right)^2 \tag{6.8}$$

For the slow-wave helical transmission line, the dispersion relation cannot be solved analytically and therefore (6.6) and (6.7) have to be solved numerically. Combining these yields:

$$x = -Z\frac{\pi f_0}{\lambda_g^2}L_{res}\frac{d\lambda_g}{df}_{|f=f_0} \tag{6.9}$$

where f stands for the frequency. The coupling of the discontinuity embedded between two identical resonators with reactance slope parameters x can be derived ([3], Chapter 8):

$$k_{j,j+1} = \frac{K_{j,j+1}}{\sqrt{x_j x_{j+1}}} \tag{6.10}$$

where equal slope parameters for the two resonators reduces to:

$$k_{j,j+1} = \frac{K_{j,j+1}}{x} \tag{6.11}$$

and can be finally expressed as:

$$k_{j,j+1} = \frac{Z}{x}\left(\sqrt{\frac{1}{|S_{12}|^2}} - \sqrt{\frac{1}{|S_{12}|^2} - 1}\right) \tag{6.12}$$

where Z is the characteristic impedance of the transmission line; implementing the resonator and the discontinuity is assumed to be symmetrical and reciprocal with zero ohmic losses and

$$\sqrt{\frac{1-|S_{11}|}{1+|S_{11}|}} = \sqrt{\frac{1}{|S_{12}|^2}} - \sqrt{\frac{1}{|S_{12}|^2} - 1} \tag{6.13}$$

by substituting x in (6.9):

$$k_{j,j+1} = -\frac{Z}{Z\dfrac{\pi f_0}{\lambda_g^2} L_{res} \dfrac{d\lambda_g}{df}\Big|_{f=f_0}}\left(\sqrt{\frac{1}{|S_{12}|^2}} - \sqrt{\frac{1}{|S_{12}|^2} - 1}\right) \tag{6.14}$$

The coupling coefficient M is derived by [6]:

$$M_{j,j+1} = \frac{f_0}{BW} k_{j,j+1} \tag{6.15}$$

In view of (6.14), it can be noticed that the characteristic impedance Z is both on the nominator and denominator, and that it cancels out. This suggests that—in the calculation of the coupling coefficient—only the reactance slope parameter and the K-inverter values are required. Figure 6.19 plots the calculated guided wavelength λ_g as a function of frequency in the range 250–350 MHz. The guided wavelength was obtained from the dispersion analysis of a single helical transmission line. The single line is assumed in a cavity with the same dimensions as in the parallel-coupled transmission lines. Figure 6.19 also plots the derivative of the guided wavelength; the

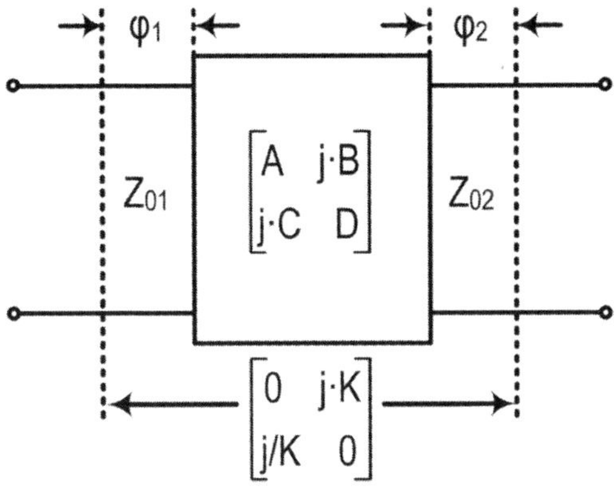

Figure 6.18 Convert a lossless two-port network in a K-inverter.

normalized *K*-inverter with respect to frequency and; finally, the normalized reactance slope parameter.

The calculated coupling coefficient as derived from (6.15) is shown in Figure 6.20(a). The coupling coefficient *M* is zero at the frequency of 336 MHz, which coincides with the frequency of the transmission zero for the second-order filter. *M* is calculated –1.58 at 291 MHz, which is approximately the same value as calculated at the center frequency of the filter. *M* is increasing almost in a linear fashion as the frequency is increasing from a value of 291 MHz up to 336 MHz. At 336 MHz, *M* is zero; above that frequency, *M* takes positive values. When introducing the frequency-dependent coupling coefficient to the coupling matrix prototype, we can obtain the new prototype response. This calculated response is shown in Figure 6.20 (b). As it can be seen, the response (as obtained from the coupling matrix prototype with frequency-dependent coupling) matches the simulated filter response. The small discrepancies are attributed to the fixed value for the transmission line characteristic impedance that is employed in HFSS in order to derive the *S*-matrix of the *K*-inverter. The frequency dispersion of the characteristic impedance of the helical transmission line is neglected, but the effect to the response of the filter is minimal. It is worth noticing that the initial linear approximation for the frequency-dependent coupling coefficient is not very distant to the one that is obtained by calculating the dispersive coupling coefficient through the *S*-parameters analysis.

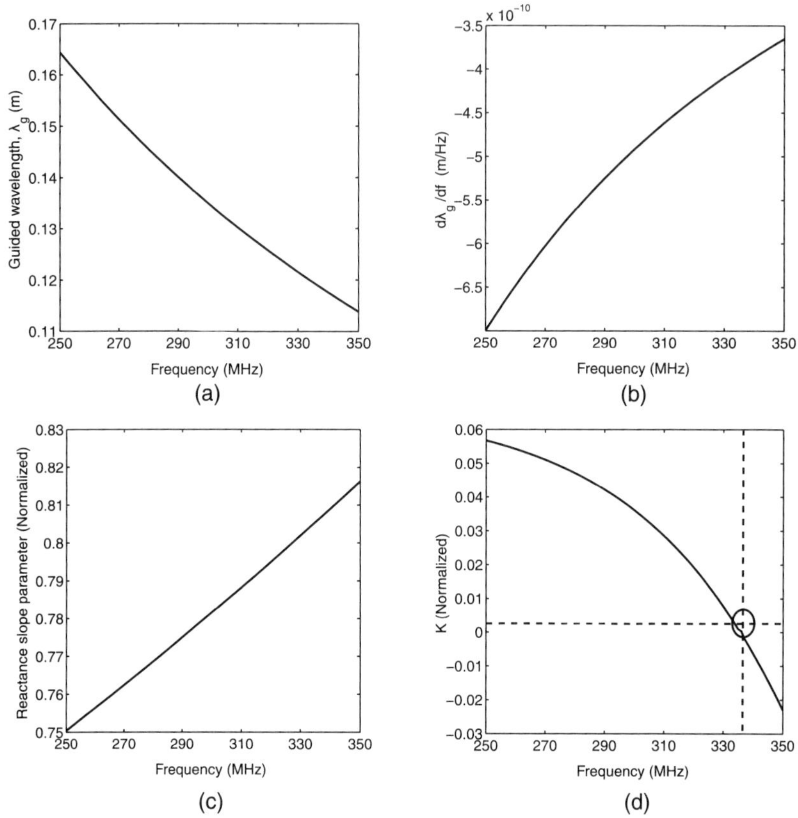

Figure 6.19 Step-by-step calculations in the derivation of the coupling coefficient *M*.

6.3.2.3 Second-Order Filter Examples

In order to illustrate the independent control of the coupling coefficient and the frequency of the transmission null, as well as to provide design guidelines, a parametric study for different values of the cavity width, *b*, and the distance of the resonator to the ground *g*, is presented (Figure 6.9). Figure 6.21 summarizes the value of the coupling coefficient and the frequency of the transmission zero for a pair of resonators with dimensions as in design 1 and varying *b* and *g*. The study reveals that negative values of the coupling coefficient (i.e., electric coupling) give rise to a transmission zero above the resonance, while positive values of the coupling coefficient give rise to a transmission zero below the resonance. Figure 6.21 also suggests that by

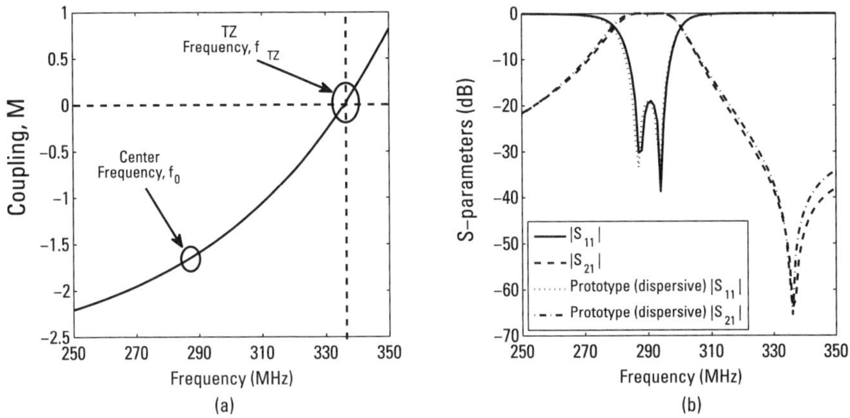

Figure 6.20 Calculated coupling coefficient *M* as a function of frequency. Prototype (*M* as a function of frequency) and simulated second-order filter response. Dimensions of the second order filter are as in Table 6.5.

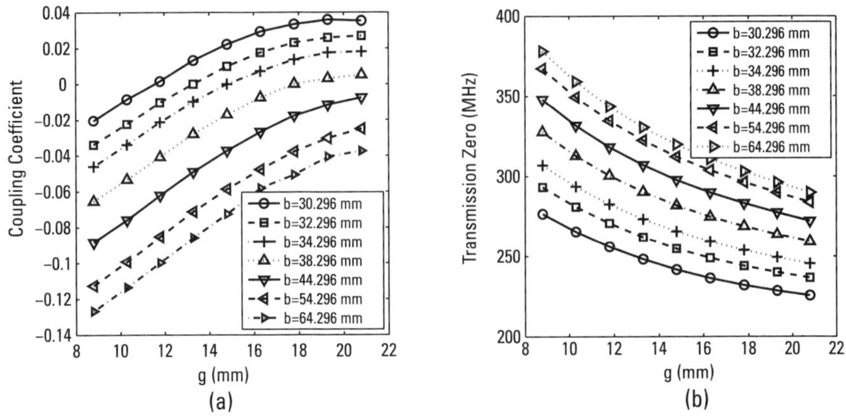

Figure 6.21 Parametric study of the coupling coefficient and the location of the transmission zero for two synchronously-tuned coupled helical resonators in an interdigital topology (see in Figure 6.9) as a function of the cavity width, *b*, and distance to ground, *g*.

simultaneously varying b and g a coupling coefficient of magnitude up to 0.05 can be achieved while selectively positioning the transmission zero in the range from 225 MHz to 375 MHz. When the further geometrical degrees of freedom are exploited—including the inter-resonator distance a_{12} and cavity height c (not shown here for brevity)—it is possible to achieve coupling coefficient magnitude in the range 0 to almost 0.1, while selectively positioning the transmission null in the range 225 to 375 MHz for a center frequency of approximately 260 MHz.

The design of interdigitally coupled helical resonator filters can then be based on the classical synthesis approach described in [1]. The required coupling matrix in the passband is initially determined. Next, an iterative procedure is followed in order to obtain initial values for the geometrical dimensions of each pair of resonators that would provide the predetermined coupling coefficient and the prescribed frequency for the TZ employing curves (such as those depicted in Figure 6.21). The input and output tapping position is designed so that the group delay of the reflection coefficient from the first resonator in isolation matches the value required by the external quality factor [1]. A second-order filter with ripple 0.07 dB, center frequency 258.8 GHz and bandwidth 7.2 MHz is employed as an example. The coupling matrix in the passband that satisfies these requirements is $k = 0.04$, and the impedances are $Q_{e1} = Q_{e2} = 28.12$. Two filters are designed employing the resonator pairs of designs 1 and 2 of Figure 6.13. The simulated responses are shown in Figure 6.22 and are in good agreement with the prototype in the passband. The transmission zeros emerge at 229 MHz and 287 MHz, respectively, as predicted by Figure 6.21.

6.3.2.4 Experimental Prototypes

To verify the above, a prototype filter with a center frequency of 244 MHz, 8-MHz bandwidth, 0.045-dB ripple, and a transmission null at 289 MHz has been designed, fabricated, and tested. Here, the design includes supporting dielectrics for mechanical stability against vibrations, as well as increased power handling (see also Chapter 7). In particular, a 2.6-mm layer of dielectric with permittivity and loss tangent has been used to cover the helices in order to improve mechanical stability and power handling. The helix and cavity dimensions are given in Table 6.6. A photograph of the fabricated prototype is shown in Figure 6.23. The helices were made in aluminium and subsequently silver-plated. No tuning elements were employed, and no other post fabrication tuning was performed. The filter's measured performance is shown in Figure 6.24, where the simulated performance is superimposed for comparison. The good agreement validates the accuracy of the full wave simulations.

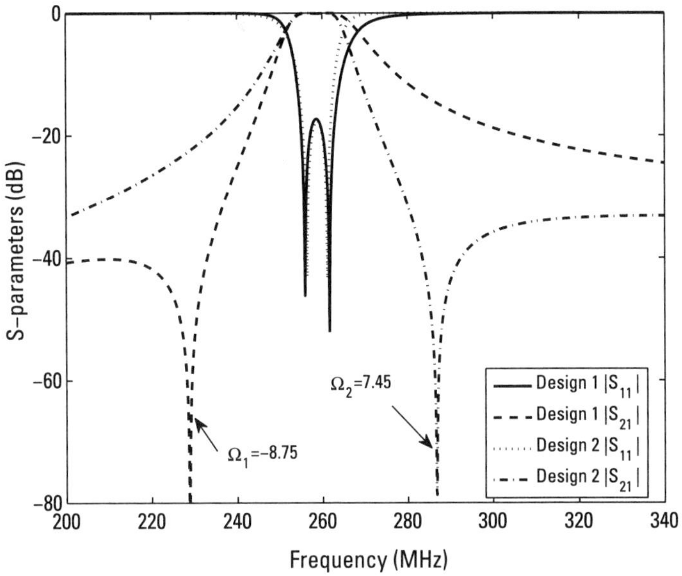

Figure 6.22 Simulated response for the two second-order filters that exhibit similar pass-band characteristics with an out-of-band TZ in the opposite side of the passband. Low-pass normalized transmission zeros at $\Omega_1 = -8.75$ and $\Omega_2 = 7.45$. (Copyright 2014 IEEE [40].)

Table 6.6
Dimensions of the Second-Order Fabricated Helical

Helix	mm(λ_0)	Cavity	mm(λ_0)
d	20.47	a_1=a_2	16.47
p	7.67	a_{12}	32.65
l	36.41	a	65.6(.065)
w	4	b	65.9(.066)
n	4.74	**c**	52.7(.053)
g	3.55		

The measured insertion loss in the passband is 0.18 dB. From this value, the unloaded quality factor of the resonators is estimated at 1,224. This value is in good agreement with the simulated unloaded quality factor for the standalone helical resonator, which is found to be 1,273. Despite the presence

Figure 6.23 Photograph of the second-order manufactured prototype filter. Dimensions are as in Table 6.6. (Copyright 2014 IEEE [40].)

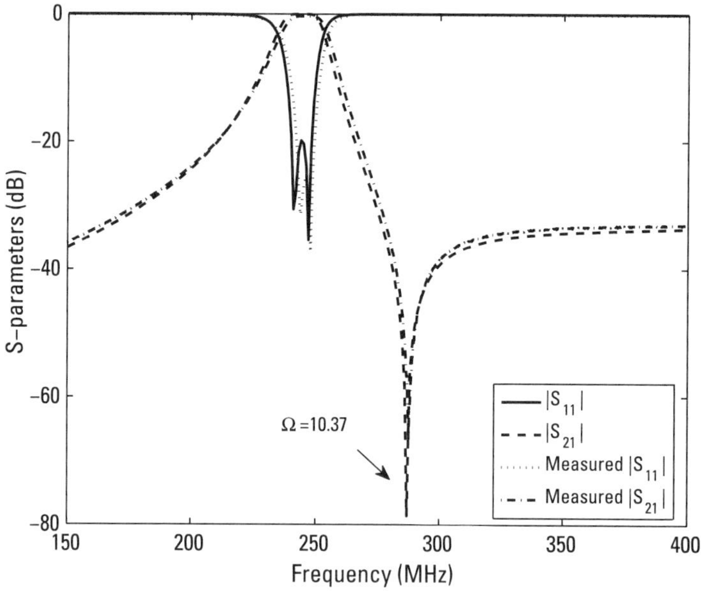

Figure 6.24 Simulated and measured results for the filter of Figure 6.23. (Copyright 2014 IEEE [40].)

of the dielectric, the measured *Q*-factor compares well with the values reported by Yu and Dokas [13]. Moreover, by virtue of the TZ, the measured response provides 30 dB improved isolation at 289 MHz, compared to the Chebyshev prototype. The overall dimensions of the filter are $a = 0.055$, $b = 0.054$, and $c = 0.043$. This proves that the proposed coupling topology allows for selective positioning of a TZ in the stopband without compromising the insertion loss and power handling capability of the filter.

Next, the application of interdigital coupling arrangement for helical resonator filters is illustrated by means of an example involving a fourth-order filter prototype. A model of the configuration for the filter is shown in Figure 6.25. The synthesis for the filter commences from the synthesis of a coupling matrix prototype, which satisfies the passband requirements. The dimensions for the helical resonators are obtained from the synthesis process (given in Chapter 5). For a more convenient implementation, the initial design considers the same width for the cavity for all of the coupled resonators, offering a uniform waveguide section for the filter. The starting dimensions for the cavity of the filter are derived from the parametric curves similar to those in Figure 6.21 for the selected initial cavity width.

The filter is designed for a center frequency of 279 MHz and a bandwidth of 16 MHz (fractional bandwidth of 5.7%). Accordingly, a fourth-order coupling matrix prototype that implements the same in-band transfer function is synthesized. The obtained values from the synthesis for the gk parameters are: $g_0 = 1$, $g_1 = 0.7028$, $g_2 = 1.1935$, $g_3 = 1.3087$, $g_4 = 0.6410$, $g_5 = 1.0964$. The normalized impedances are R1 = R4 = 1.4228, and the symmetric coupling matrix is:

$$M = \begin{bmatrix} 0 & -1.0918 & 0 & 0 \\ -1.0918 & 0 & -0.8001 & 0 \\ 0 & -0.8001 & 0 & -1.0918 \\ 0 & 0 & -1.0918 & 00 \end{bmatrix} \tag{6.16}$$

The simulated response of the filter is shown in Figure 6.26, where the response of the prototype is superimposed. The filter exhibits three TZs, of which two are placed below and the third is placed above the passband of the filter. The frequency of the transmission zeros is mainly controlled by the width of the filter's cavity. The TZ placed above the passband follows the trends as observed for the parallel-coupled resonators. When the cavity width

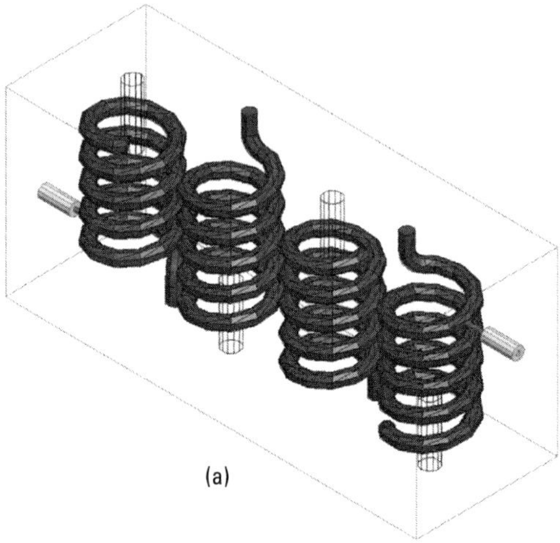

(a)

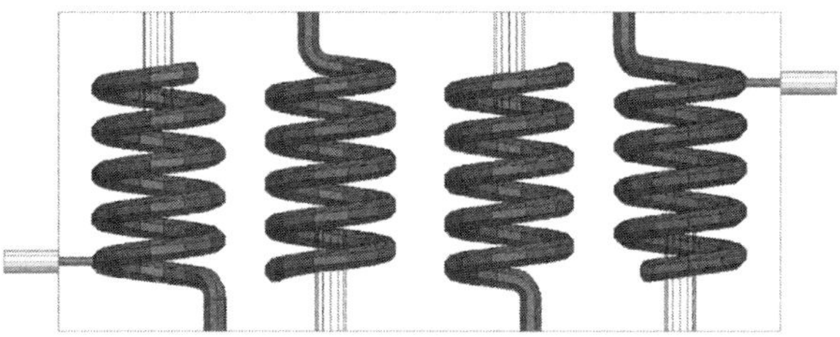

(b)

Figure 6.25 Fourth-order interdigital filter. In-line filter topology.

is increased, the frequency is shifted towards higher values. The two TZs at the lower side of the passband are attributed to cross-couplings between the first and third resonators and first and fourth resonators. This cross-coupling is feasible because the coupling between the resonators is significantly increased due to the wider cavity width, *b*. The dimensions of the filter are given in Table 6.7.

The filter has been fabricated and tested. A photograph of the fabricated prototype is shown in Figure 6.27. The simulated and measured results are superimposed in Figure 6.28, validating the predicted transmission zeros. Additional tuning effort is required to tune for equiripple passband response.

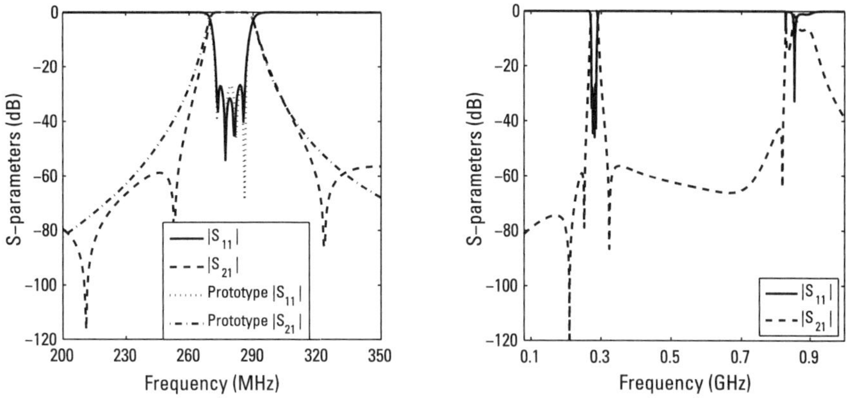

(a) (b)

Figure 6.26 Fourth-order interdigital filter. Simulated response (a) close-to-band, (b) out-of-band.

Table 6.7
Dimensions of the Fourth-Order Helical Filter; Helix and Cavity

Helix	**mm(λ_0)**	**Cavity**	**mm(λ_0)**
d	20.47	a_1=a_2	18.20
p	7.67	a_{12}=a_{34}	31.23
l	36.41	a_{23}	32.55
w	4	b	45
n	4.74	c	56.7
g	5.48		

Figure 6.27 Photo of the fourth-order interdigital filter prototype.

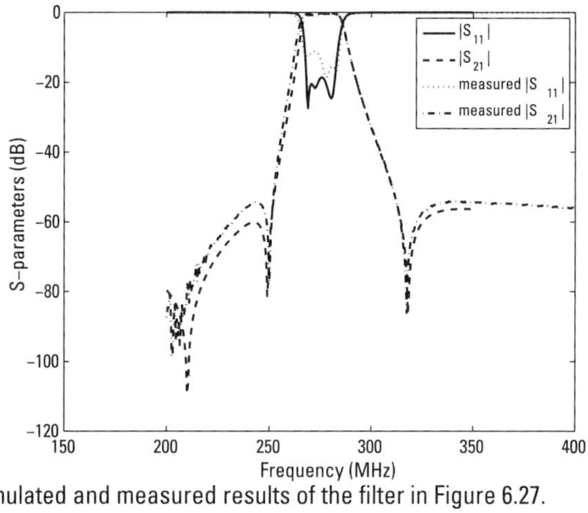

Figure 6.28 Simulated and measured results of the filter in Figure 6.27.

References

[1] Hong, J., and M. Lancaster, *Microstrip Filters for RF/Microwave Applications*, New Jersey: Wiley-Interscience, 2001.

[2] Cameron, R., C. Kudsia, and R. Mansour, *Microwave Filters for Communication Systems: Fundamentals Design and Applications*, New Jersey: Wiley-Interscience, 2007.

[3] Matthaei, G., L. Young, and E. Jones, *Microwave Filters Impedance-Matching Networks and Coupling Structures*, New York: McGraw-Hill, 1964.

[4] Hunter, I., *Theory and Design of Microwave Filters*, London, UK: the Institution of Electrical Engineers, 2001.

[5] Darlington, S., "A history of network synthesis and filter theory for circuits composed of resistors inductors and capacitors," *IEEE Transactions on Circuits and Systems I: Fundamental Theory and Applications*, Vol. 46, No. 1, pp. 4–13, 1999.

[6] Atia, A. E, A. E. Williams and R. Newcomb, "Narrow-band multiple-coupled cavity synthesis," *IEEE Transactions on Circuits and Systems*, Vol. 21, No. 5, p. 649–655, 1974.

[7] Cameron, R., "General coupling matrix synthesis methods for Chebyshev filtering functions," *IEEE Transactions on Microwave Theory and Techniques*, Vol. 47, No. 4, p. 433–442, 1999.

[8] Cameron, R. J, " Advanced coupling matrix synthesis techniques for microwave filters," *IEEE Transactions on Microwave Theory and Techniques*, Vol. 51, No. 1, p. 1–10, 2003.

[9] Wang, H., and Q. Chu, "An inline coaxial quasi-elliptic filter with controllable mixed electric and magnetic coupling," *IEEE Transactions on Microwave Theory and Techniques*, Vol. 57, No. 3, pp. 667–673, 2009.

[10] El Sabbagh, M., K. Zaki, H. Yao, and M. Yu, "Full-wave analysis of coupling between combline resonators and its application to combline filters with canonical configurations," *IEEE Transactions on Microwave Theory and Techniques*, Vol. 49, No. 12, pp. 2348–2393, 2001.

[11] Levy, R., "Filters with single transmission zeros at real or imaginary frequencies," *IEEE Transactions on Microwave Theory and Techniques*, Vol. 24, p. 172–181, 1976.

[12] S. Amari, "On the maximum number of finite transmission zeros of coupled resonator filters with a given topology," *IEEE Microwave and Guided Wave Letters*, Vol. 9, No. 9, p. 354–356, 1999.

[13] Yu, M., and V. Van Dokas, "High performance helical resonator filters," in 34th European Microwave Conference, Amsterdam, 2004.

[14] Levy, R., and C. La Jolla, "New cascaded trisections with resonant cross-couplings (CTR sections) applied to the design of optimal filters," in *IEEE MTT-S International Microwave Symposium Digest*, 2004.

[15] Levy, R., "Direct synthesis of cascaded quadruplet (CQ) filters," *IEEE Transactions on Microwave Theory and Techniques*, Vol. 43, No. 12, p. 2940–2945, 1995.

[16] Levy, R., and P. Petre, "Design of CT and CQ filters using approximation and optimization," *IEEE Transactions on Microwave Theory and Techniques*, Vol. 49, No. 1, pp. 12–25, 2001.

[17] Thomas, J., "Cross-coupling in coaxial cavity filters-a tutorial overview," *IEEE Transactions on Microwave Theory and Technique*, Vol. 51, No. 4, pp. 1368–1376, 2003.

[18] Rhodes, J., and R. Cameron, "General extracted pole synthesis technique with applications to low-loss TE/011/mode filters," *IEEE Transactions on Microwave Theory and Techniques*, Vol. 28, pp. 1018–1028, 1980.

[19] Wang, Y., and M. Yu, "True inline cross-coupled coaxial cavity filters," *IEEE Transactions on Microwave Theory and Techniques*, Vol. 57, No. 12, p. 2958–2965, 2009.

[20] Menzel, W., F. Alessandri, A. Plattner, and J. Bornemann, "Planar integrated waveguide diplexer for low-loss millimeter-wave applications," in 27th European Microwave Conference, Jerusalem, 1997.

[21] Amari, S., and J. Bornemann, "Using frequency-dependent coupling to generate finite attenuationpoles in direct-coupled resonator bandpass filters," *IEEE Microwave and Guided Wave Letters*, Vol. 9, No. 10, pp. 404–406, 1999.

[22] Amari, S., J. Bornemann, W. Menzel, and F. Alessandri, "Diplexer design using pre-synthesized waveguide filters with strongly dispersive inverters," in *IEEE MTT-S International Microwave Symposium Digest*, 2001.

[23] Hong, J., and M. Lancaster, "Couplings of microstrip square open-loop resonators for cross-coupled planar microwave filters," *IEEE Transactions on Microwave Theory and Technique*, Vol. 44, No. 11, pp. 2099–2109, 1996.

[24] Menzel, W., and M. Berry, "Quasi-lumped suspended stripline filters with adjustable transmission zeroes," in *IEEE MTT-S International Microwave Symposium Digest*, 2004.

[25] Ma, M., J. Ma, K. Yeo and M. Do, "A compact size coupling controllable filter with separate electric and magnetic coupling paths," *IEEE Transactions on Microwave Theory and Techniques*, Vol. 54, No. 3, pp. 1113–1125, 2006.

[26] Chu, Q., and H. Wang, "A compact open-loop filter with mixed electric and magnetic coupling," *IEEE Transactions on Microwave Theory and Techniques*, Vol. 56, No. 2, p. 431–439, 2008.

[27] Levy, R., and K. K. Andersen, "An optimal low loss HF diplexer using helical resonators," in *IEEE MTT-S International Microwave Symposium Digest*, 1992.

[28] Shen, W., L. Wu, X. Sun, W. Yin and J. Mao, "Novel substrate integrated waveguide filters with mixed cross coupling (MCC)," *IEEE Microwave and Wireless Components Letters*, Vol. 19, No. 11, pp. 701–703, 2009.

[29] Cohn, S., "Direct-coupled resonator filters," *Proceedings of the IRE*, Vol. 2, pp. 187–196, 1957.

[30] Hwang, H., and S. Yun, "The design of bandpass filters considering frequency dependence of inverters," *Microwave Journal*, Vol. 45, No. 9, pp. 154–163, 2002.

[31] Vanin, F., D. Schmitt and R. Levy, " Dimensional synthesis for wide-band waveguide filters and diplexers," *IEEE Transactions on Microwave Theory and Techniques*, Vol. 52, No. 11, 2004.

[32] Soto, P., E. Tarin, V. Boria, C. Vicente, J. Gil, et al, "Accurate synthesis and design of wideband and inhomogeneous inductive waveguide filters," *IEEE Transactions on Microwave Theory and Techniques*, Vol. 58, No. 8, p. 2220–2230, 2010.

[33] Zaki, K., W. Meng, H. M. Lee, and A. E. Atia, "Synthesis of multi-coupled resonator filters with frequency-dependent couplings," in *IEEE MTT-S International Microwave Symposium Digest*, 2010.

[34] Zverev, A., *Handbook of Filter Synthesis*, New York: Wiley, 1967.

[35] Crino, V., and S. Kosmopoulos, "Metop—the SRA diplexer design-technical report," *Space Engineering S.p.A*, Rome, 1999.

[36] Cohn, S., "Parallel-coupled transmission-line-resonator filters," *IRE Transactions on Microwave Theory and Techniques*, Vol. 6, No. 2, pp. 223–231, 1958.

[37] Pozar, D., *Microwave Engineering*, Hoboken, NJ: John Wilely & Sons, 2005.

[38] Zysman, G. I., and A. K. Johnson, "Coupled transmission line networks in an inhomogeneous dielectric medium," *IEEE Transactions on Microwave Theory Techniques*, Vol. 17, pp. 753–759, 1969.

[39] Tsai, C. M., S. Y. Lee and H. M. Lee, "Transmission-line filters with capacitively loaded coupled lines," *IEEE Transactions on Microwave Theory and Techniques*, Vol. 51, No. 5, pp. 1517–1524, 2003.

[40] Doumanis, E., G. Gousettis, and S. Kosmopoulos, "Inline interdigital pseudo-elliptic helical resonator filters," *IEEE Microwave and Wireless Compnents Letters,* Vol. 21, No. 8, pp. 400–402, 2011.

[41] Levy, R., "A generalized design technique for practical distributed reciprocal ladder networks," *IEEE Transactions on Microwave Theory and Techniques*, Vol. 21, p. 519–526, 1973.

7

Helical Resonator Filters with High Power-Handling Capability for Space Applications

7.1 Introduction

Improving the power-handling capability of microwave filters is becoming an increasingly relevant challenge for RF/microwave engineers and scientists. Base station, radar, and satellite applications are typical applications that are in need of such high-power components in order to improve spectral efficiency, range, and quality of service. Microwave breakdown phenomena occurring in these components can compromise the operation of the communications systems and even damage equipment [1]. The microwave breakdown phenomena are divided mainly in two categories: the ionization breakdown [1, 2], and the multipaction breakdown [3]. The *multipaction* is a radio frequency (RF) breakdown mechanism related to vacuum conditions, whereas the *ionization* is a gas (such as air) or other insulator breakdown phenomenon. Eventually, the applications of cellular base-stations and radar communication systems are typically subject to ionization breakdown, whereas satellite components are primarily subject to multipaction breakdown phenomena.

This chapter discusses high-power considerations in RF passive components for space applications. Starting with a brief top-level description on multipactor [3–5], some critical information on the design and test of RF components for multipaction-free performance as defined in the European Space Agency (ESA) standard [4] is outlined. The power handling cabability of helical resonator technology is assessed by means of an example. The difficulties emerging in the employment of this technology for high-power space applications is illustrated. A technique that increases the power handling without significantly affecting the quality factor, size, or mass is presented together with validation from an experimental test campaign.

7.2 Multipactor

Multipactor is a microwave breakdown discharge that occurs in vacuum conditions caused by the formation of an electron avalanche. It takes place when the mean free path of electrons is longer than the gap between the walls guiding the RF power. In this case, the free electrons that are accelerated by the applied RF voltage can travel towards the walls with a minimum risk of collisions in this path. A few electrons initially present are accelerated across the gap between the walls due to the presence of an electric field. Some of the electrons may collide with one of the surfaces. The electrons can be absorbed, reflected, or can extract secondary electrons from the surface. Secondary electrons are produced at a rate that depends on the impact energy and angle of incidence of the primary incident electrons, as well as the material properties. The latter is summarized by the secondary emission yield (SEY) coefficient of the surface. The SEY is quantitatively described by means of the secondary electron emission coefficient (SEEC or δ) as the ratio of secondary electrons emitted by the surface divided by the number of incident primary electrons.

If the conditions (e.g., frequency, geometry) are such that when the electrons hit a surface the field is reversed, then the electrons are accelerated towards the opposite surface, thus creating an electron resonance. If the SEY is sufficiently higher than unity, then the number of oscillating elements could be rapidly increased, creating an electron avalanche. This finally results in the appearance of an electron plasma, which reduces the output power of the component and increases its return loss. If the multipactor conditions are sustained for a long period of time, out-gassing of the walls increases the pressure inside the component and a corona discharge can develop, which can result in a total destruction of the component.

Strictly, the condition for multipaction to occur is that the gap size should be a multiple of the half-cycle of the applied RF voltage in order to satisfy the conditions of electron resonance [6, 7]. In general, the frequency gap, or $f \times d$, product becomes relevant in multipactor investigation. In practice, this product is linked with the maximum and the minimum breakdown field strengths at various excitation frequencies. Hatch and Williams in [5] have presented a set of experimental results investigating multipaction. The experimental setup consisted of two flat, circular discharge electrodes in a chamber under low-pressure conditions. Aluminum, silver-plated copper, and copper electrodes were used. Multipaction susceptibility zone charts were developed, where the breakdown field strength was plotted against frequency. As a result of this study, semiempirical equations that estimate the breakdown region were produced. A historical perspective on multipaction is presented by Vaughan in [3], where the phenomenon is identified in a number of applications. An analysis of two parallel-plate surfaces is conducted for normal and oblique angles of incidence for the electrons. Typical curves of versus impact voltage are given for normal and almost 55° angles of incidence. In this work, the electron velocity is associated to an equivalent voltage.

Multipactor inherently depends on the secondary emission properties of the surfaces involved. The release of secondary electrons when high energy impact electrons collide with the metallic walls of the device has been investigated by Vaughan [8]. Mathematical expressions for the SEY as a function of impact voltage and direction are presented, and a formula is proposed. The shape of the curve for the secondary emission ratio (δ) is somewhat universal, but the parameters vary over a considerable range. More recently, it is most common that the SEY curves are plotted against the impact energy as opposed to the impact equivalent voltage used by Vaughan [4].

The multipactor discharge on a dielectric in the context of ceramic RF windows, where a theory of a single surface multipactor is presented, has been studied in [9]. A parallel to the dielectric surface RF field and a constant DC field normal to the surface is assumed. The formulas of Vaughan are adopted for the SEEC (δ). As stated in [8], the dependence of on-impact energy is qualitatively similar for most materials. In [10], an improved model of a single-surface multipactor is presented. Two novel aspects were introduced from [9]: namely, the beam loading on the external RF by the multipaction discharge, and the evolution of the charging static field normal to the surface. A positive static field appears because of the electrons leaving the dielectric surface, thus charging it positively. The evolution of the DC electric field depends on the dielectric charging. This charging field is caused by the positive charges that are assumed to spread uniformly over the dielectric surface. It is concluded that the surface charging of the dielectric may lead to the saturation of multipactor discharge. In the same year, a historical review on multipactor discharge on metals and dielectrics was produced in [11].

The multipactor effects in dielectric loaded waveguides have been recently addressed [12, 13]. The geometry of the problem involves a parallel plate waveguide that is partially filled with a dielectric layer. It differs from the single surface multipactor because this problem is a two-surface problem, and the orientation of the RF electric field is normal to the surfaces. A theory based on the classical Hatch and Williams model [5] has been developed; this theory includes the presence of the thin dielectric. The time evolution of the multipaction discharge is studied. In [12], susceptibility zone charts are given for different metallic and dielectric materials. The peak voltage across the gap (V_0) is plotted against the frequency-gap product $f \times d$. It is concluded that the multipactor is partially eliminated when coating the waveguide for some V_0-$f \times d$ points of the susceptibility chart where multipactor occurred in the empty waveguide. The effect of the dielectric layer included in the model is that it generates a time-varying static electric field.

In more recent work [14], the analysis is extended to include the space charge effects and considers multiple effective electrons [15]. In order to model a multipactor, the use of effective electrons is employed in such a way that the real number of electrons to be tracked remains constant throughout the calculation; however, each electron represents a larger (or smaller) number of electrons, which depends on the SEEC at each impact. A modified version of Vaughan's formula is employed. The value of δ is calculated by means of realistic SEY functions that include the effect of reflected electrons for low-impact energies [15]. The results obtained with the inclusion of space-charge effects demonstrate that the number of electrons in the case of two metallic surfaces reaches a maximum value after a few RF cycles. As soon as the population of electrons reaches a significant number, the electron avalanche affects the effective electrons, modifying their trajectories. This results in loss of the synchronization with the RF field. The consequence is low-impact energy collisions so that the metallic plates absorb electrons.

The appearance of a time-varying static electric field in the empty gap is shown to extinguish with time the multipaction discharge for particular operating conditions [14]. This is attributed to the time-varying static electric field; when the static field becomes intense enough, the positively charged dielectric material attracts effective electrons towards the dielectric layer. A number of electrons that collide with the opposite metallic surface have lower impact energies due to the static field. This results in electrons being absorbed in the collisions ($\delta < 1$). Thus, a fewer number of electrons collide against the dielectric, fewer electrons are emitted, and the static field does not increase as much. The DC field finally reaches a constant value, whereas the population of electrons is decreased until it finally disappears as electrons are absorbed by the metallic surface.

7.3 Space Qualifications Procedures

In this section, critical information in the standard for multipaction design and test as proposed by the European Cooperation for Space Standardization (ECSS) is outlined. According to this standard, the generation of multipaction depends strongly on the following factors [4][1]:

- Vacuum condition;
- Applied RF voltage;
- Frequency of operation in conjunction with geometry of RF components;
- Surface conditions.

Possible ways to prevent the occurrence of this phenomenon are given in [16]:

- Control of the frequency-gap product (large gap size, gap size below cutoff);
- Dielectric filling;
- Pressurization;
- Magnetic or dc bias;
- Reduction of the surface potential by treatment of materials.

A summary of advantages and disadvantages associated with these multipactor mitigation techniques is presented in [16]. Technological trade-offs should therefore be considered in light of other important factors (i.e., cost, complexity, applicability, etc.).

1. http://multipactor.esa.int/ECSS-E-20-01A.pdf, last accessed on 05/09/2010.

7.3.1 Verification

The process of verification of a component with respect to multipaction performance involves the conformance to the required margins. The margins are applied according to the classification of the component type. The classification of the component types (according to the materials employed in the construction) is given in [4] and reproduced in Table 7.1. The applicable margins for each type of component are given in Tables 7.2, 7.3, and 7.4 for Type 1, Type 2, and Type 3 components, respectively. The route to conformance is illustrated in the flow diagram given (reproduced from [4], Section 4) and is repeated in Figure 7.1. The relevant tests as depicted in Figure 7.1 are defined as follows:

- *Design margin.* Theoretically computed margin between the specified power handling of the component and the result of an analysis after the analysis uncertainty has been subtracted.
- *Qualification test.* Testing performed on a single flight standard unit to establish that a suitable margin exists in the design and build standard. The suitable margin is the qualification margin.
- *Batch acceptance test.* Test performed on a sample from each batch of flight units to verify that the units conform to the acceptance requirements.
- *Unit acceptance test.* Testing carried out on each flight standard unit to verify that the unit conforms to the acceptance requirements.

Table 7.1
Classification of Component Type

Type	Characteristics
1	The RF paths are entirely metallic (with known secondary electron emission properties), or are metallic with a non-organic surface treatment that increases the multipaction threshold. Note that this does not preclude the use of coated plastics or CFRP, (provided that only metal surfaces are subjected to the RF fields). The components are well-vented.
2	The RF paths contain or can contain dielectrics or other materials for which the multipaction performance is well-defined. The components are well-vented.
3	Any components not classified as Type 1 or Type 2.

From: ESA ECSS [4].

The conformity for analysis only should meet the following requirements (Analysis conformity, Figure 7.1):

1. There is a proven heritage of similar qualified designs.

Table 7.2
Margins Applicable to Type 1 Components

Route	Analysis (dB)	Qualification Test (dB)	Batch Acceptance Test (dB)	Unit Acceptance Test (dB)
1	8	—	—	—
2	—	6	—	—
3	—	—	4	—
4	—	—	—	3

From: ESA ECSS [4].

Table 7.3
Margins Applicable to Type 2 Components

Route	Analysis (dB)	Qualification Test (dB)	Batch Acceptance Test (dB)	Unit Acceptance Test (dB)
1	10	—	—	—
2	—	6	—	—
3	—	—	4	—
4	—	—	—	3

From: ESA ECSS [4].

Table 7.4
Margins Applicable to Type 3 Components

Route	Analysis Margin	Qualification Test Margin (dB)	Batch Acceptance Test Margin dB)	Unit Acceptance Test Margin (dB)
1	12	—	—	—
2	—	10	—	—
3	—	—	6	—
4	—	—	—	4

From: ESA ECSS [4].

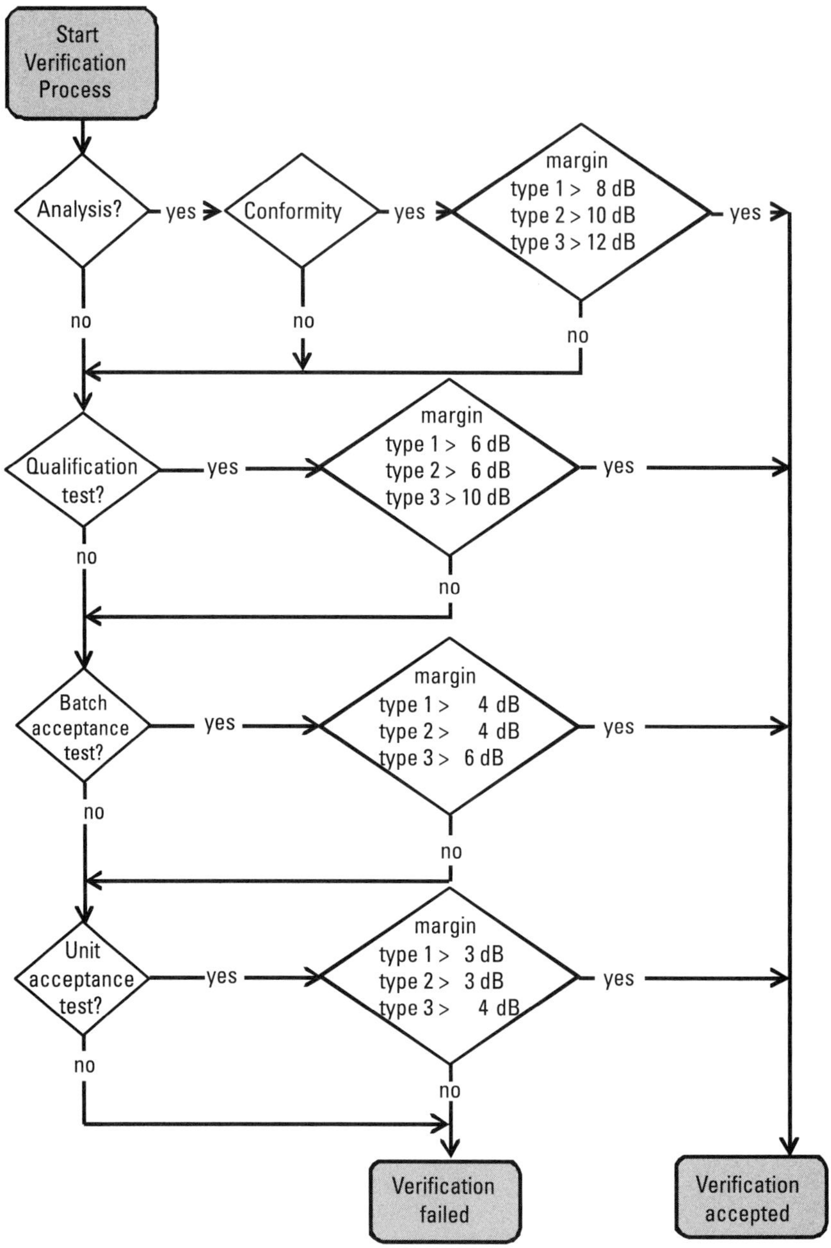

Figure 7.1 Routes to conformance for single carrier. (Copyright 2014 ESA ECSS [4].)

2. The component has a geometry that allows accurate field calculations to be performed with high confidence.
3. The multipaction-critical areas of the component have commonality with an existing design that has established the correlation between analysis and test.

7.3.2 Analysis

The scope of this approach in the multipaction analysis is to calculate the threshold input power beyond which multipaction phenomena are possible to occur. According to the ECSS standard [4], the first step is to identify the most critical areas for multipaction inside of the component (passive component, for example, filter). In particular, full-wave electromagnetic simulations are employed in order to identify all regions where strong electric fields and narrow gaps exist. For each critical gap, the frequency-gap product and voltage are estimated. The *voltage magnification factor* (VMF) associated with the device under test is extracted as the ratio of the peak voltage at the most critical region divided by the input voltage. The predicted multipaction threshold is then calculated using available *multipactor susceptibility charts* that plot the multipactor threshold voltage against the frequency-gap product for each material [4]. A typical multipaction susceptibility chart is shown in Figure 7.2. A software tool to assist in this procedure is available [17].

The aforementioned procedure exploits available susceptibility charts that were originally extracted for parallel plate waveguide geometries for a given range of metals (alodine, silver, gold, copper, and aluminum). Recommendations for margins in the cases of singe and multicarrier signals are also provided. It is noted that such approaches may lead to conservative estimations in cases where the geometry of the most critical gap differs significantly from a parallel plate waveguide. Commercial tools for multipactor prediction exploiting advanced numerical techniques that model the electron cloud development in the presence of the electromagnetic field have emerged [18, 19] and provide promising routes for reducing margins.

7.3.3 Methods of Detection

The multipactor detection methods as reported in [4, Section 7] can be divided into two categories:

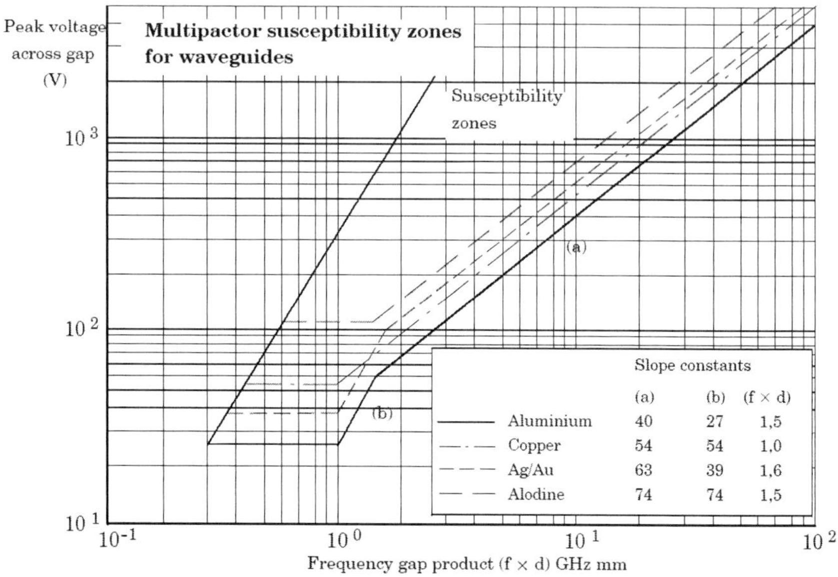

Figure 7.2 Multipaction susceptibility zones for parallel plates for aluminium, copper, silver, and alodine. (Copyright 2014 ESA ECSS [4].)

1. Global detection methods:
    - Close to carrier noise;
    - Phase noise;
    - Harmonic noise;
    - Microwave nulling.
2. Local methods:
    - Optical;
    - Electron probe.

Local detection methods are used close to the point of the actual discharge, while most global detection methods monitor noise levels in the components response. Microwave nulling refers to a specific interferometric technique where the signals incident and scattered from the device are superimposed with weights such that they cancel out. Multipactor then is manifested as a disruption in the null signal. A review on the detection methods is given in [4, 6], along with block diagrams of the most commonly used measurement setups. It is important to note that, due to the high sensitivity of the detection methods, a number (usually three) of methods are used simultaneously. As a general rule, multipaction is inferred when at least two of the methods agree in the indication. The aforementioned techniques are sufficiently sensitive so that the onset of multipactor can be detected and the power level can be reduced before any discharge or other damaging effect on the component occurs.

7.4 Helical Resonators for Downlink Filters

7.4.1 Power Handling Analysis

A second-order helical resonator filter operating in the ultrahigh frequency (UHF) band is employed as an example to demonstrate the power handling limitations due to multipaction [20]. The design guidelines for the filter are given in Chapter 6. Table 7.5 gives the dimensions of the helices and the cavity of the filter that are employed in this example. The simulated response of the filter is shown in Figure 7.3. The center frequency of the filter is at 300 MHz and the fractional bandwidth is 5.3%. Based on the critical region identification steps described earlier, an input at the filter of a single carrier with time average power of 1W is assumed. Employing a commercial full-wave electromagnetic simulator, a field analysis of the structure is performed. The electric field distribution inside the filter at the center frequency of 300 MHz is calculated and shown in Figure 7.4. The field is stronger closer to the helix and gets weaker as it approaches the cavity wall. For this example, the most critical region is marked in Figure 7.4. The peak voltage calculated by the line

Table 7.5
Second-Order Filter—Helix and Cavity Dimension

Helix	**mm(λ_0)**	**Cavity**	**mm(λ_0)**
d	20.47	$a_1=a_2$	16.47
p	7.67	a_{12}	32.65
l	36.41(.036)	a	65.59 (.065)
w	4	b	65.89 (.065)
n	4.74	c	52.71 (.053)
g	3.55		

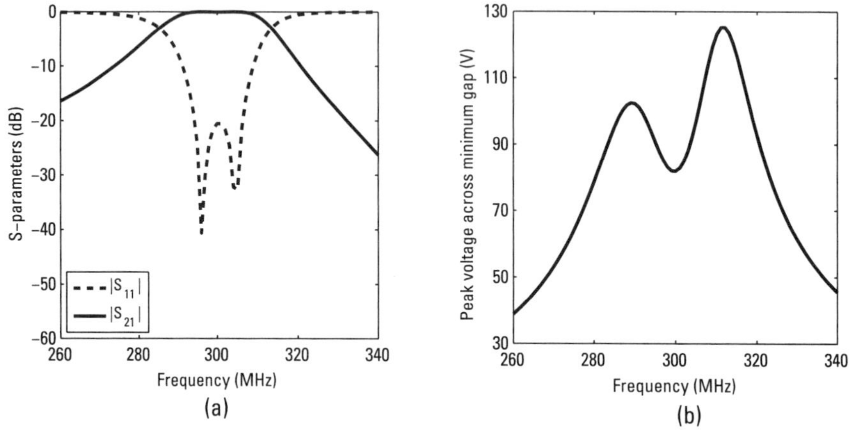

Figure 7.3 (a) Simulated response of the second-order filter, and (b) calculated peak voltage across most critical region over frequency (time average power of 1W).

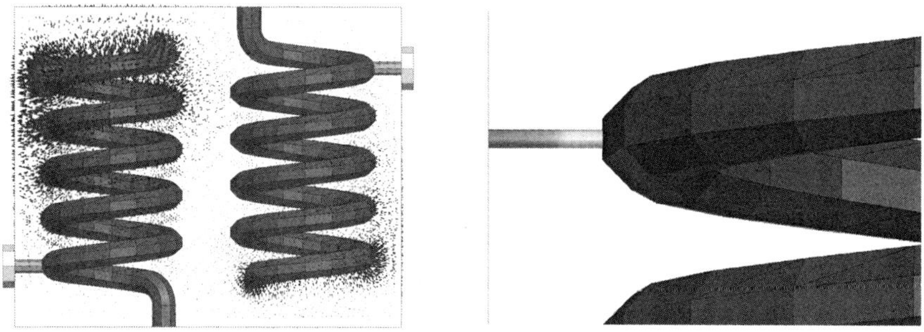

(a) (b)

Figure 7.4 Electric field intensity (V/m) distribution at the center frequency of the filter (most critical region identification). Calculation of the voltage across the identified gap.

integral of the electric field across the straight path of length 4.3 mm depicted in Figure 7.4 is found to be equal to 124V at approximately 300 MHz.

According to the above analysis, the frequency-gap product is 1.352 GHz × mm. Referring to the susceptibility zone boundaries for silver, the filter is therefore liable to potential onset of multipaction for voltages beyond the breakdown value of approximately 72.8V [4]. The input power that produces this breakdown voltage is found to be 0.35W. Space verification standards require 8-dB margins from the theoretically calculated power level, so that this filter is qualified for 17.5 dBm. It is quite clear that, because the operating frequency is very low (i.e., in the UHF) and the dimensions of the helical filter are very small compared to the wavelength, this filter configuration deals with significant difficulties as to how much input power it can handle. Additionally, it is impractical to apply the large-gap approach [16] because, for helical resonator filters, the cavity dimensions are typically close to the dimensions of the helices.

According to [16, Section 20.6.5] prevention of multipaction is feasible by dielectric filling. By filling the empty spaces of the discharge region with solid dielectric, the mean free path of electrons is reduced, shifting the multipactor threshold. However, this would result in a major decrease in the resonator unloaded *Q*, and is therefore considered to be a great disadvantage [21]. In our example, this decrease is quantified for space qualified Rexolite1422 (dielectric permittivity $\varepsilon_r = 2.53$, and loss tangent $tan\delta = 3 \times 10^{-4}$), and given in Table 7.6. For comparison purposes, Table 7.6 also gives the unloaded *Q* of the initial air-filled resonator. A significantly greater insertion loss would be the immediate consequence, particularly for filters of higher order, as well as greater total mass and cost.

7.4.2 Dielectric Covers

An investigation of multipactor discharges in a parallel plate waveguide structure where a layer of dielectric covers one of the metallic plates has been recently presented in [12, 14]. In [12], it was demonstrated that the inclusion of dielectric extends the multipaction-free region for low values in the $f \times d$ product axis of the susceptibility chart. Furthermore, in [14], it was shown that the use of the dielectric layer extinguishes the multipaction discharge with time due to the build-up of a static field produced by the charging of the dielectric layer. Following this approach, it is here proposed to cover all the

Table 7.6

Unloaded Quality Factor for the Helical Resonator with Dimensions as in Table 7-5 in the Cases of an Unfilled Cavity, a Cavity Fully Loaded with Rexolite 1422 and the Selected Solution of a Rexolite 1422 cover.

Case	Simulated Q_0 (f_0)	Measured Q_0
Empty cavity	1340 (298 MHz)	—
Cavity fully loaded with Rexolite	780 (188 MHz)	—
Rexolite cover	1273 (245 MHz)	1238

metallic parts of the resonating helices by a dielectric layer. In this way, both the mass and the insertion loss of the filter can be maintained at the levels of the air-filled filter, adding further mechanical stability. As it can be seen in Table 7.6, the unloaded Q of the dielectric loaded resonator is only slightly decreased from 1340 to 1273 when a layer of 2.6 mm of Rexolite 1422 is used to cover the resonators, demonstrating that the effect on the insertion loss is minimized. For arbitrary geometries, like a helical filter and mixed material cavities, the lack of tools to predict the breakdown onset, limits the theoretical power handling estimation, which is therefore tested experimentally according to the verification routes described in [4, Section 4].

The assessment of the filter's power-handling capability is based on a test campaign involving a fabricated prototype. Figure 7.5 shows a photograph of the fabricated prototype. The helices were made in aluminum and then silver-plated in order to reduce the insertion loss[2] and improve the multipaction-free filter operation [16]. A layer of 2.6 mm of Rexolite 1422 was used to cover the helices. We note that due to fabrication limitations, the lower end of the helix (lower than the TNC[3] connector in Figure 7.6) was not covered by the dielectric. The simulated and measured reflection and

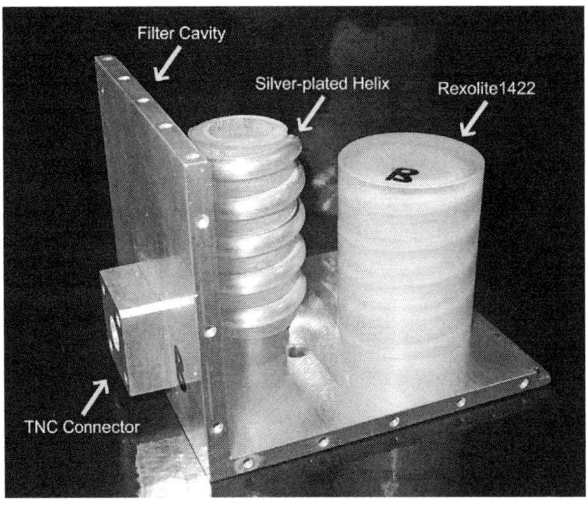

Figure 7.5 Fabricated prototype of the second-order helical resonator filter. (Copyright 2014 IEEE [20].)

2. The conductivity of silver is higher than that of aluminum.
3. RF coaxial connector.

transmission of the filter are superimposed in Figure 7.7(a) for comparison. The in-band insertion loss is shown in Figure 7.7(b). The electric field distribution inside the filter at the center frequency is shown in Figure 7.8. Good agreement between the simulated and measured results is observed. From the measured insertion loss in the passband, the unloaded *Q* of the resonator is found to be 1238, in good agreement with the simulated one which is 1273.

A multipaction measurement of this prototype was performed in David Florida Laboratory of the Canadian Space Agency.[4] The multipaction measurement conducted for the experimental verification of this work is based on a combination of three different global detection methods, namely the forward/backward nulling, the third harmonic response, and the noise close to carrier. The methods rely on the effect that multipaction has on the return loss, the third harmonic response and the sideband noise, respectively. Monitoring all three signals simultaneously minimizes the possibility of false detection. The onset of multipaction is manifested as a simultaneous disturbance on at least two of these signals [4].

Multipaction test results are presented in Figure 7.9. The test frequencies were chosen at the center frequency of the filter, and at the lower band

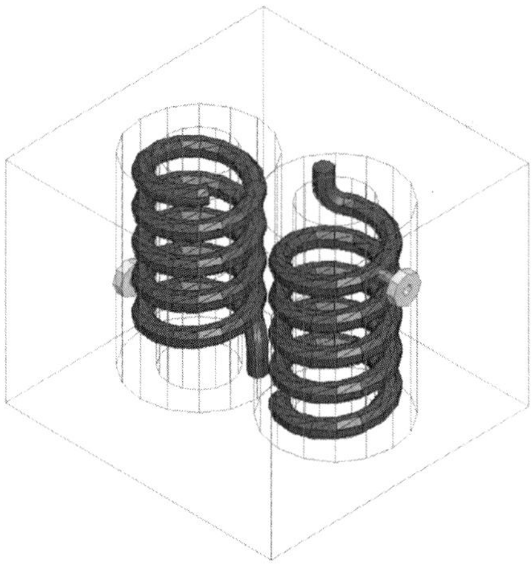

Figure 7.6 This prototype second-order filter model configuration illustrates dielectric support frame and dielectric cover for the helices.

4. http://www.asc-csa.gc.ca/eng/dfl/radio.asp#7.

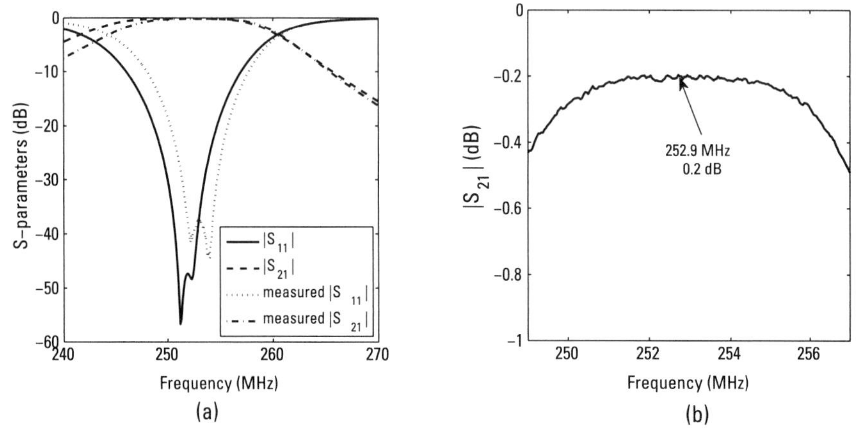

Figure 7.7 Simulated and measured filter response (second-order filter with Rexolite1422 cover). (Copyright 2014 IEEE [20].)

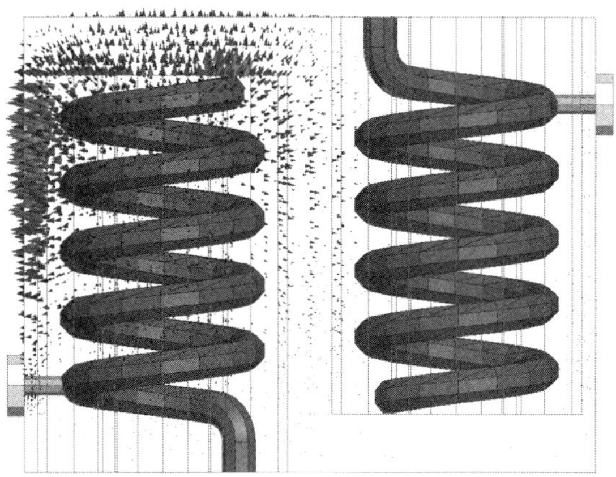

Figure 7.8 Electric field intensity distribution at the center frequency of the filter (most critical region identification).

edge [4, 22]. According to the protocol described in [4, Section 6.5], if the component contains resonant features, it should be tested at the center frequency and the band edges. Pulse testing has been used in the measurements. It can be applied in the case of components operating in either a continuous wave (CW) or a pulsed mode. For units that experience CW excitation,

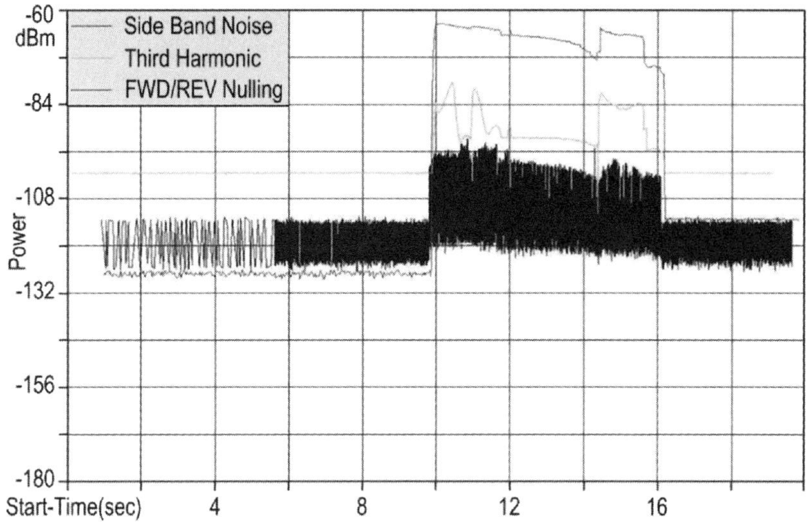

(a)

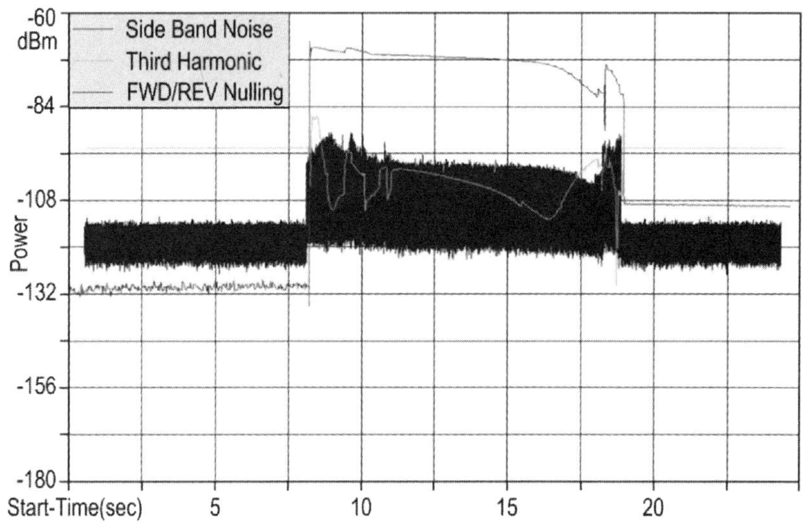

(b)

Figure 7.9 Multipaction test results for helical resonator filter prototype shown in Figure 7.7. (a) Multipaction results at 250.8 MHz with a peak power of 90W on a 30% duty cycle. (b) Multipaction results at 252.9 MHz with a peak power of 100W on a 30% duty cycle. (Copyright 2014 IEEE [20].)

pulsed testing can be used to achieve maximum test power, while keeping the mean power within the specifications of the unit and permitting the use of lower cost test equipment. The pulsed profiles are calculated based on the maximum power to be applied to the component and are summarized in Table 7.7.

The measurement procedure includes an initial measurement with a starting input power of 10W and subsequent tests increasing the power at a step of 10W. The measured design has shown that multipaction is affecting the filter's performance at a power level of 90W at 250.8 MHz. The forward/reverse nulling power level jumps from a value of less than –120 dBm to a value of almost –60 dBm after about 10 seconds of pulsed operation, as shown in Figure 7.9(a). This indicates an imbalance in the null at this power level, which was not observed in any previous steps. A disturbance was observed in the other two methods as well. Similar phenomena occur at a power level of 100W at 252.9 MHz as shown in Figure 7.9(b). In this case, the disturbance occurs approximately eight seconds after the initiation of the test. At an input power of 300W a complete breakdown occurs in the measurement and all power is reflected. We note that visual observation of the filter following the test showed no signs of damage anywhere within the filter.

7.5 Summary and Conclusions

The power-handling aspects of satellite filters are primarily determined by multipactor effects, which are associated with a coupled electron cloud/elec-

Table 7.7
Calculated Pulse Profiles Based on Maximum Power on the Device Under Test

Step	Peak Power (W)	On Duty (μsec)	Duty Cycle (%)	Off Duty (μsec)	T_{mod} (usec)	F_{mod} (kHz)
1	300	10	30	23.3	33.3	30
2	600	10	15	56.7	66.7	15
3	900	10	10	90.0	100.0	10
4	1200	10	7.5	123.3	133.3	7.5
5	1500	10	6.0	156.7	166.7	6
6	2400	10	3.8	256.7	266.7	3.75

tromagnetic resonance. Strict procedures apply for space qualification relating to multipactor. The testing relies on a range of global or local observation techniques. When it comes to helical resonator filters—despite their advantageous characteristics in terms of volume and quality factor—power-handling capacity can be significantly compromised. A technique to improve power handling without substantially sacrificing the mass and volume or *Q*-factor exploits thin covers of a space-qualified dielectric covers around the helices. This technique is here demonstrated by means of an example involving a second-order filter; multipaction analysis carried out for an example of an uncoated helical resonator filter has given a very low power level (0.35W) for the multipaction onset. Such power levels are too low for most practical space applications of distributed filters. Based on the above and coupled with the very high cost associated with such measurement, the verification through measurement for the uncoated case was assumed considered redundant.

The conducted study provides evidence that, by virtue of the dielectric coating, multipaction was not detected in the most critical region identified for the uncoated filter. This can be attributed either to the discharge switching off with time before the effects can be experimentally detected [14] or to shifting of the multipaction risk zone to higher $f \times d$ values, extending the risk-free zone to include the region under consideration [15]. Further numerical analysis predicted an input power threshold of about 51 dBm in the uncoated region between the lower uncovered part of the helix and the cavity wall of the fabricated prototype. This calculated value is comparable to the experimentally obtained value of about 50 dBm, and is an indication that multipaction may have been initiated in that part of the filter.

Additionally, a possible alternative to this design approach would be the use of a thin dielectric covering to the metallic walls of the filter cavity instead of the metallic parts of the helices. This solution, however, is associated with increased mechanical instability. This is due to the absence of any dielectric support frame for the helices. In conclusion, a multipaction analysis for helical resonator filters is presented and shows that the power handling capabilities are severely limited. A second-order UHF filter was used as an example, and it was shown that (according to verification standards) it qualifies for power input up to 17.8 dBm. In order to improve the power handling capability, it was proposed to introduce dielectric covers of the resonators. A prototype employing a space-qualified dielectric has been fabricated and experimentally tested. The measurement results showed initiation of multipaction at input power of 90W. Space verification standards require 4 dB margins from the measured multipaction onset power, so that the power handling capability of the coated filter after the verification margins is 45.5 dBm. This corresponds to more than 27 dB improvement compared to the uncoated case with reduction in the unloaded *Q*-factor limited to about 5%.

References

[1] Yu, M., "Power-handling capability for RF filters," *IEEE Microwave Magazine*, Vol. 8, No. 5, p. 88–97, 2007.

[2] Woo, R., "Final report on RF voltage breakdown in coaxial transmission," Jet Propulsion Lab, California, 1970.

[3] Rodney, J., and M. Vaughan, "Multipactor," *IEEE Trans. Electron Devices*, Vol. 35, No. 7, pp. 1172–1180, 1988.

[4] ECSS Organization, "Space engineering, multipaction design and test," *European Cooperation for Space Standardization*, ESA ESTEC, Noordijk, 2003.

[5] Hatch, A., and H. Williams, "The secondary electron resonance mechanism of low-pressure high-frequency gas breakdown," *Journal of Applied Physics*, Vol. 25, p. 417, 1954.

[6] Tang, W., C. Kudsia, C. Dev, and O. Cambridge, "Multipactor breakdown and passive intermodulation in microwave equipment for satellite applications," in *IEEE Military Communications Conference (MILCOM'90)*, Monterey, CA, 1990.

[7] Kudsia, C., R. Cameron, W. Tang, and O. Cambridge, "Innovations in microwave filters and multiplexing networks for communications satellite systems," *IEEE Transactions on Microwave Theory and Techniques*, Vol. 40, No. 6, pp. 1133–1149, 1992.

[8] Rodney, J., and M. Vaughan, "A new formula for secondary emission yield," *IEEE Transactions on Electron Devices*, Vol. 36, No. 9, p. 1963–1967, 1989.

[9] Kishek, R., and Y. Lau, "Multipactor discharge on a dielectric," *Physical Review Letters*, Vol. 80, No. 1, pp. 193–196, 1998.

[10] Ang, L., Y. Lau, R. Kishek, and R. Gilgenbach, "Power deposited on a dielectric by multipactor," *IEEE Transactions on Plasma Science*, Vol. 26, No. 3, pp. 290–295, 1998.

[11] Kishek, R., Y. Lau, L. Ang, A. Valfells, and R. Gilgenbach, "Multipactor discharge on metals and dielectrics: Historical review and recent theories," *Physics of Plasmas*, Vol. 5, p. 2120, 1998.

[12] Torregrosa, G., A. Coves, A. San Blas, A. Pérez, C. Vicente, B. Gimeno, et al, "Analysis of multipactor effect in dielectric-loaded waveguides," in *Proc. 5th International Workshop on Multipactor, Corona and Passive Intermodulation in Space RF hardware*, ESA ESTEC, Noordijk, Netherlands, 2005.

[13] Torregrosa, G., A. Coves, C. Vicente, A. Perez, B. Gimeno, et al, "Time evolution of an electron discharge in a parallel-plate dielectric-loaded waveguide," *IEEE Electron Device Letters,* Vol. 27, No. 7, pp. 619–621, 2006.

[14] Coves, A., G. Torregrosa-Penalva, C. Vicente, B. Gimeno and V. Boria, "Multipactor Discharges in Parallel-Plate Dielectric-Loaded Waveguides Including Space-Charge Effects," *IEEE Transactions on Electron Devices,* Vol. 55, No. 9, pp. 2505–2511, 2008.

[15] Vicente, C., M. Mattes, C. Wolk, B. Mottet, H. Hartnagel, et al, "Multipactor breakdown prediction in rectangular waveguide based components," in *IEEE MTT-S International Microwave Symposium,* Long Beach, CA, 2005.

[16] Cameron, R., C. Kudsia, and R. Mansour, *Microwave Filters for Communication Systems: Fundamentals, Design and Applications,* Hoboken, NJ, Wiley-Interscience, 2007.

[17] E. S. Agency, *ECSS Multipactor Tool.*

[18] SPARK3D/FEST3D, Aurora Software and Testing, Valencia, 2014.

[19] "CST Particle Studio, Computer Simulation Technology (CST), [Online]. Available at: https://www.cst.com/Products/CSTPS/TechnicalSpecification. Last accessed 10 July 2014.

[20] Doumanis, E., G. Goussetis, W. Steffe, D. Maiarelli, and S. Kosmopoulos, "Helical Resonator Filters With Improved Power Handling Capabilities for Space Applications," *IEEE Microwave Components Letters,* Vol. 20, No. 11, pp. 598–600, 2010.

[21] Hunter, I., L. Billonet, and B. G. P. Jarry, "Microwave filters-applications and technology," *IEEE Transactions on Microwave Theory and Techniques,* Vol. 50, No. 3, p. 794–805, 2002.

[22] Sivadas, A., M. Yu, and R. Cameron, "A simplified analysis for high power microwave bandpass filter structures," in *IEEE MTT-S International Microwave Symposium Digest,* Vol. 3, pp. 1771–1774, 2000.

8

Summary

RF and microwave filters for space applications share some specific features and requirements that stem from their application domain. In this book, we have attempted to illustrate the role of filters in satellite systems, and in particular, in the communications payload. Through this review, the challenges associated with RF and microwave filter design for space are identified. Fundamental technology challenges in filters for satellite systems are summarized, and a rich bibliography is provided that will benefit the interested reader. The specific focus of this book has been to illustrate helical resonator filter technology for space applications, with the view of making it approachable to design engineers and researchers alike.

Helical resonator filters offer significant advantages, particularly at the lower part of the RF/microwave spectrum. This is due to the favorable trade-off that they offer in terms of *Q*-factor and volume/mass budget. A contemporary and comprehensive treatment of the technology targeting high-performance filters for satellite systems is provided. Based on modern CAD tools, both physical insight and practical design methodologies have been presented. Practical techniques that allow one to synthesize the physical dimensions of the filter from the elements of the coupling matrix have been provided, leading to design-cycle time and cost savings. Numerous practical, real-life applications of helical resonator filters and real-life hardware implementations of the technology have been used to illustrate the technology. State-of-the-art research outcome that can help the advanced reader succeed in the design of filters with even the most stringent specifications is included throughout the book.

Significant original contributions to helical resonator filters are presented throughout the book. The synthesis of the physical dimensions of helical resonators is treated based on the dispersion properties of coaxial helical transmission lines. A novel, interdigital coupling topology for helical resonator filters is considered; its use in realizing selectively located TZs in in-line filters is demonstrated with performance metrics both qualitatively and quantitatively characterized. A specific aspect associated with the usage of helical resonator technology for high-power space applications is discussed. A top-level description of high-power phenomena in satellite systems is presented. A multipaction analysis for helical resonator filters is conducted; it reveals severe limitations in power handling capabilities due to multipaction. It is proposed to introduce dielectric covers of the resonators in order to improve the power handling capability. Measured results demonstrate an improvement in the power handling of the filter with a minimum reduction in the unloaded *Q*-factor.

Although dating back a few decades—and while helical transmission lines have also been treated in the context of travelling wave tubes (e.g., [1]), the topic of helical resonator filters remains a vibrant one [1–6]. Several challenges remain open for future engineers to address. For example, while this book was being finalized, new technologies for reconfiguring helical resonator filters were being proposed [2]. In parallel, research efforts to improve power handling capabilities by virtue of modulating the radius of the helix along its length are being placed [3]. With the advent of additive manufacturing technologies, topics that could emerge in the future include their use in the 3-D printing of helical filters. Manufacturing developments could also enable the use of helical resonator filters at higher frequencies, possibly realized in multilayer configurations [4]. New materials and space qualification aspects could further provide additional routes for helical resonator filter improvement. Modeling multipactor phenomena in helical resonator filters is also a topic of increasing interest [5]. It is hoped that this book summarizes the present state of the art on helical resonators with emphasis on space applications, and will therefore also prove to be a good reference for future researchers in this area.

References

[1] Pierce, J., "Travelling-Wave Tubes," *The Bell System Technical Journal*, Vol. 29, No. 1, pp. 1–59, 1950.

[2] Psychogiou, D., and D. Peroulis, "Tunable VHF Miniaturized Helical Filters," *IEEE Transactions on Microwave Theory and Techniques*, Vol. 62, No. 2, pp. 282–289, 2014.

[3] Doumanis, E., G. Goussetis, S. Kosmopoulos, and M. Guglielmi, "Conical-Shaped Helical Resonators for Improved Multipaction-Free Power Thresholds," in *European Space Agency Microwave Technology and Techniques Workshop*, Nordwijk, 2012.

[4] Heo, K., J. Lim, J. Mun, and S. Hwang, "Characterization and Wideband Modeling of Miniaturized LTCC Helical Inductors," *IEEE Microwave and Wireless Components Letters*, Vol. 17, No. 3, pp. 160–162, 2007.

[5] "SPARK3D/FEST3D," Aurora Software and Testing, Valencia, 2014.

[6] Zhao, Z.-Y., P.-H. Li, K.-L. Cheng, and W.-Q. Cao, "High Power VHF Frequency-Hopping Filters with High Suppression of Second Harmonic," *Progress in Electromagnetics Research Letters*, Vol. 20, pp. 119–128, 2011.

List of Abbreviations

3-D	Three-dimensional
BW	Bandwidth
CAD	Computer-aided design
CFRP	Carbon fiber reinforced plastic
CST	Computer simulation technology
CTE	Coefficient of thermal expansion
CW	Continuous wave
D/C	Down converter
DC	Direct current
DUT	Device under test
ECSS	European cooperation space standardization
EIRP	Effective isotropic radiated power
EM	Electromagnetic energy
ESA	European Space Agency
FBW	Fractional bandwidth
FDD	Frequency division duplexing
GD	Group delay
GPS	Global Positioning System

HF	High frequency
HFSS	High-frequency structure simulator
HPA	High-power amplifier
I/O	Input/output
IF	Intermediate frequency
IL	Insertion loss
IM	Intermodulation
IMUX	Input multiplexer
IPT	Initial performance test
LNA	Low-noise amplifier
NASA	National Aeronautics and Space Administration
OMUX	Output multiplexer
PIM	Passive intermodulation
RF	Radio frequency
RL	Reflection/return loss
SEEC	Secondary electron emission coefficient
SEY	Secondary emission yield
SMA	Subminiature version A (RF connector)
TE	Transverse electric
TM	Transverse magnetic
TL	Transverse lossl
TNC	Threaded Neill-Concelman (RF connector)
TZ	Transmission zero
VHF	Very high frequency
VMF	Voltage magnification factor
U/C	Upconverter
UHF	Ultrahigh frequency

About the Authors

Savvas A. Kosmopoulos received both his B.A.Sc. (1978) and M.A.Sc. (1981) in electrical engineering from the University of Ottawa, Canada. In 1980, he joined the Bell Northern Research laboratories in Ottawa, where he has been involved in research activities on digital communication networking. In 1987, he was awarded a Ph.D. in electrical engineering from the University of Thessaloniki, Greece, where he was elected lecturer in October of the same year. From 1987 to 1989, he worked as a postdoctoral researcher in the Laboratory for Electromagnetics and Microwaves, Department of Electrical Engineering, University of Ottawa.

Since 1989 he has worked at Space Engineering S.p.A., Rome, Italy, where he is currently a senior engineer in the RF and payload group. During the last two decades, Dr. Kosmopoulos has been actively involved in various ESA projects dealing with the RF/IF linear and nonlinear analyses of several payload and/or ground segment systems (e.g., the ARTEMIS satellite project, the METEOSAT II Generation system, the COSMO SKYMED system, and recently, the TDP#5 payload on-board the ALPHASAT spacecraft). These activities have permitted his involvement in the direct design and realization of several related RFs, microwaves and millimeter-waves, small signals, and high-power components and subsystems (e.g., frequency converters and filters, diplexers, multiplexers). Dr. Kosmopoulos is the author of over 60 technical papers published in various international refereed journals, as well as in the proceedings of many international conferences. He is a member of the Institute of Electrical and Electronic Engineers and the Technical Chamber of Greece.

George Goussetis received his B.A. in electrical and computer engineering from the National Technical University of Athens, Greece, in 1998, and his Ph.D. from the University of Westminster, London, in 2002. In 2002, he also received his B.Sc. in physics (first class) from University College London (UCL), UK.

In 1998, he joined Space Engineering, S.p.A. Rome, Italy, as an RF engineer. In 1999, he joined the Wireless Communications Research Group, University of Westminster, UK, as a research assistant. Between 2002 and 2006, he served as a senior research fellow at Loughborough University, UK. He was a lecturer (assistant professor) with Heriot-Watt University, Edinburgh, UK, between 2006 and 2009, and a reader (associate professor) with Queen's University Belfast, UK, between 2009 and 2013. In 2013, he joined Heriot-Watt as a reader and was promoted to professor in 2014. He has authored and coauthored over 200 peer-reviewed papers, three book chapters, and two patents. His research interests include the modeling and design of microwave filters, frequency-selective surfaces and periodic structures, leaky wave antennas, microwave sensing and curing, and numerical techniques for electromagnetics.

Dr. Goussetis held a research fellowship from the Onassis foundation in 2001, a research fellowship from the U.K. Royal Academy of Engineering between 2006 and 2011, and a European Marie-Curie experienced researcher fellowship in 2011–12. In 2010, he was a visiting professor at UPCT, Spain. He is the corecipient of the 2011 European Space Agency Young Engineer of the Year prize, the 2011 EuCAP Best Student Paper Prize, and the 2012 EuCAP Best Antenna Theory Paper Prize. He serves as associate editor to *IEEE Antennas and Wireless Propagation Letters.*

Efstratios Doumanis graduated from the Electrical and Computer Engineering School at the Democritus University of Thrace, Greece, in April 2005, and received his Ph.D. from Queen's University of Belfast, U.K, in 2010. In 2003, he joined KFKI Research Institute for Particle and Nuclear Physics, Department of Space Technology, Budapest, Hungary, as a student engineer. Between 2008 and 2009, he was with the RF/Microwaves and Payload Group of Space Engineering S.p.A, (now ASTRIUM) Rome, Italy, as an RF engineer. In 2010, he joined the High Frequency Electronic Circuits (HFEC) Group at Queen's University of Belfast, UK, as a research assistant. In 2012, he joined Alcatel Lucent, the RF-Front End and Antenna Group, where he is currently a member of the technical staff.

Mr. Doumanis was a corecipient of the Best Student Paper award in the 2011 European Conference on Antennas and Propagation (EuCAP), Rome, Italy, and corecipient of the 2011 European Space Agency Young Engineer of the Year prize.

Index

Recent Titles in the Artech House Space Technology and Applications Series

Bruce R. Elbert, Series Editor

Business Strategies for Satellite Systems, D. K. Sachdev

Filter Design for Satellite Communications: Helical Resonator Technology, Efstratios Doumanis, George Goussetis, and Savvas Kosmopoulos

Gigahertz and Terahertz Technologies for Broadband Communications, Terry Edwards

Ground Segment and Earth Station Handbook, Bruce R. Elbert

Introduction to Satellite Communication, Third Edition, Bruce R. Elbert

Laser Space Communications, David G. Aviv

Low Earth Orbital Satellites for Personal Communication Networks, Abbas Jamalipour

Mobile Satellite Communications, Shingo Ohmori, et al.

Radio Interferometry and Satellite Tracking, Seiichiro Kawase

Satellite Broadcast Systems Engineering, Jorge Matos Gómez

The Satellite Communication Applications Handbook, Second Edition, Bruce R. Elbert

Satellite Communications Fundamentals, Jules E. Kadish and Thomas W. R. East

Satellite Communications Network Design and Analysis, Kenneth Y. Jo

The Satellite Communication Ground Segment and Earth Station Handbook, Second Edition, Bruce Elbert

Understanding GPS: Principles and Applications, Second Edition, Elliott D. Kaplan and Christopher J. Hegarty, editors